Ultimate Reality

Ultimate Reality

Ken Moseley

SHEPHEARD-WALWYN (PUBLISHERS) LTD

First published in 2006 by
Shepheard-Walwyn (Publishers) Ltd
Suite 604, The Chandlery
50 Westminster Bridge Road
London SE1 7QY

British Library Cataloguing in Publication Data
A catalogue record of this book
is available from the British Library

ISBN-13: 978-0-85683-237-6
ISBN-10: 0-85683-237-5

Typeset by Alacrity,
Banwell Castle, Weston-super-Mare
Printed through Print Solutions, Wallington, Surrey

Contents

Acknowledgements

I AM VERY GRATEFUL to Dolly for unwittingly having set me off on this quest, and for the many later discussions with her and her husband Michael Nanson as the book progressed. I am indebted to Dr Peter Bowman for his careful scrutiny of the manuscript and invaluable suggestions. I would also like to thank my publisher, Anthony Werner, for his contribution and Jean Desebrock for her skilled work on the text and compilation of the index. Finally, the work could not have been completed without the constant support and encouragement of my wife Lucia.

To my wife Lucia

This journey through the cosmos
was planned in the hope that it
would answer questions.
A particular wish was to know
how the cosmos works and how
it is that we can be aware of
the phenomenon.

The pilgrimage was enjoyed
immensely and it is now hoped
that those who read this
travelogue will find the
adventure stimulating.

1

Appearances are Deceptive

THIS ADVENTURE began when my wife and I were enjoying drinks with friends after a game of golf. The catalyst was an old oak table, of which there were a number in the club-house. The feel of their organic surfaces was somehow pleasing while sitting and talking on many issues. Their concealed past intrigued; at least that is how I like to relate with their subtle timelessness.

On this particular occasion I tapped the much-polished table top and questioned whether the others had any idea of the ceaseless activity going on below the apparently solid surface. I explained that it was mainly empty space, the hardness being an illusion caused by the rapid motion of the subatomic particles.

One of our friends didn't believe me. To her the table was solid reality and my talk of minute particles the illusion. To a certain extent I agreed, since I do not share the philosophy of those who state that only mind is real, all else being illusion. On the other hand, I do accept what modern science has revealed about the structure of matter. So, to avoid further confusion, I promised to write her a letter setting down more clearly what I was trying to say.

Later, when considering that letter, I realised that there was more to be said than just describing the atom. There was something much more; something underlying the atoms that makes them move the way they do. The trouble was that, like the rest of mankind, I did not really know what this was. I only had the faintest glimpse of it. This lack of real understanding did not just apply to atoms but to all other aspects of the universe. There seemed nowhere to go to gain access to this knowledge: even though the surface detail had been extensively explored, what lay behind was unknown territory. At that point I put the task of writing the letter to one side in order to think about the issue.

I soon realised that our whole planet is a mass of activity that can ultimately be reduced to tiny atoms. The horrendous tsunami disaster that killed over 200,000 people in December 2004 very effectively demonstrated that Earth's crust is not as *terra firma* as we think. Indeed, despite appearances, everything within this expanding universe is of the same nature. Consequently, my considerations grew well beyond the original topic of conversation. The promised letter has matured into a book and continuing related discussions with my friends provided sufficient excuse for me not to send it.

It has since been agreed that, despite its disasters, the world in which we live is a wonderful place. For the inquiring mind there is no end to the fascinating things that come along and arouse one's wonder and interest. A lifetime is too short to encompass everything. But perhaps that does not matter really. As the years pass what becomes more important is not so much the particular details in themselves but the question of what lies behind it all.

This book recounts my own journey searching for this elusive truth. As the field of mankind's knowledge of the world has grown ever vaster, a 'single unified theory', a 'theory of everything', has become increasingly remote and seems to have defeated the minds of the greatest scientists of our age. But maybe their defeat was not because of their inadequacy either individually or collectively.* Perhaps it was because they were using a method that could produce, at best, only partial success.

So what does 'everything' include? When I first considered this question it became clear that a survey of our present knowledge would need to take place. But this was not all. To be totally comprehensive an investigation of the unique attributes of knowingness and consciousness had to be included.

That the seemingly solid desk at which I sat writing was actually made up of minute particles consisting primarily of rapid motion in empty space was one thing. But how is it possible for us to know this? That also needs to be explained. What is this marvellous human gift of understanding that enables us to penetrate into the very interior of these particles? How am I able to know that the sun, hiding up there behind the clouds, the life force of our entire solar system, does not rise in the east and set in the west, as it appears to do from Earth's own rotation? Even though very few have actually been up into space to see this directly, there does seem to be within all our minds the capacity to grasp the reality of the solar system, the galaxy, even the whole universe.

To understand the nature of knowingness is surely as important as understanding how our planet and its species evolved. This awareness is one of the main characteristics that differentiate us from the other inhabitants of our planet. Recent discoveries have shown that our genes are surprisingly similar to those of other species, but in our appreciation of the world the difference is enormous.

*This was written before the journal *Science* listed 125 'critical knowledge gaps'. See Vol 209, 1 July 2005 (www.sciencemag.org).

It is clear that there are two layers of human existence. There is the outer aspect that includes eating, drinking and even mankind's sports, and an inner realm, which includes philosophical discussion, thought and questioning not only the external manifestations of the world but also its meaning and purpose. A 'theory of everything' surely needs to encompass both aspects.

One way to seek the fundamental cause is to use the method of analysis to break things down into their component parts. Our understanding, for example, has been greatly aided by the discovery and general acceptance of the fact that the objects of the universe are built with atoms. But to ask what the substance of these atoms is produces a question that is quite difficult to answer. And what about patterns of thought? They certainly exist and can be described in the same way as tangible things. They must, therefore, be formed of some kind of substance.

It became obvious to me that the analytical method of delving deeper and deeper into smaller and smaller parts was not the way forward. A rational understanding of mind and consciousness could only emerge from an exploration of the cosmic whole. Separate parts surely cannot be treated in isolation. A real appreciation can only result from an investigation of everything. How else can we determine how substance or cause evolved?

But where to start? An alternative to analytical or reductionist enquiry is to seek for fundamental causes by looking back into the past. The starting point would then be the beginning of time. The view now held by many is that the universe came into being with what is generally referred to as a 'big bang', an explosive event that caused the simplest atoms to form and subsequently galaxies and stars like our own sun. From within these, as they passed through their life cycle, more complex atoms formed. Few would dispute that Earth and all life on it, including mankind, are structured from these tiny bits of stardust. But serious doubts remain as to whether this explanation fully accounts for our physical bodies, thought patterns and conscious awareness.

Stephen Hawking is one of the most knowledgeable exponents of the big bang theory and other aspects of cosmology. His standing was enhanced when his book *A Brief History of Time* became a long-time bestseller, with sales running into millions. It obviously captured the hearts and minds of those who 'wanted to know'. However, the work later gained the reputation for being one of the most bought but least read books on the bookshelves of the world. This was certainly not because of any lack of clarity and wit on the part of the author, but does suggest that this approach to seeking answers to the question of the origin of the universe was not one which fully satisfied many readers.

Hawking is the Lucian Professor of Mathematics at Cambridge, a chair previously held by Sir Isaac Newton. Three hundred years separate these two exceptional men's careers yet, despite all the effort expended at this and other prestigious seats of learning, the nature of the cosmos is still shrouded in mystery.

Through his theory of universal gravity Isaac Newton showed that the apple falling to the ground, the movement of the tides, the moon, planet Earth and all heavenly bodies were governed by a single force. With the mathematical tools he developed, these movements could be described and predicted with great precision. But these achievements came at a price.

As a means of communication, mathematics is virtually incomprehensible to people who are not conversant with this subject. In consequence, the exploratory work of physicists and mathematicians has become isolated. Their pride in past successes has led to an arrogance, which assumes this method to be superior and treats with disdain other forms of thought. These specialists have lost touch with the masses and so are largely forgotten. This lack of communication must restrict the advancement of our species.

To facilitate their exacting work, physicists use their abstract algebraic language, models and other such aids to create mental images in their efforts to explain behaviour patterns. The branch of physics called quantum mechanics exemplifies the situation. It originated as a means of solving some of the difficulties involved in describing phenomena at the atomic level, but some of its consequences apply to cosmological issues. It is a mathematical theory of dynamic systems in which the variables that specify the behaviour of the system are determined by the action of mathematical operators on wave functions. It has many features that seem to be contrary to common sense, and a reputation of being so abstract that it can only ever be used as a tool that works but is never fully understood. Therefore, intuitive reasoning needs to be introduced to guide and direct this adventure towards 'ultimate reality'.

Endeavouring to get the balance right between intuition and mathematically based logic is not a new aim of mankind. The early Greek philosopher Zeno, a contemporary of Parmenides and Socrates, gives an example which indicates that the problem is at least two and a half thousand years old.

Zeno proposed that in any race between two athletes in which the slower contestant is given an adequate start, the faster mover would not be able to overtake him. He argued that the pursuer must first arrive at the point from which the pursued started. As he does so, the slower one must always have moved ahead.

The traditional illustration of this is a race between the swift–footed Achilles and a tortoise. Achilles is assumed to run ten times faster than the tortoise so the tortoise is given a 100 ft start. To overtake the tortoise Achilles must first get to where the tortoise started but in that time the tortoise has moved on 10 ft. Achilles runs the 10 ft, but in the meantime the tortoise has moved on a foot. So it goes on ad infinitum and Achilles can never reach the tortoise.

The mathematical reasoning seems infallible, yet it is quite clear who would be the winner in an actual race. The conundrum illustrates a human dilemma. There is a need for a balance between intuition and the rigorous language of mathematics. We must surely make use of both in a complementary way.

This journey within the cosmos will aim to show that astronomy, physics, chemistry, biochemistry and genetics are sufficiently advanced to allow an intuitively acceptable understanding of the cosmos to be established. The first step will be to review what is known about the world through these scientific disciplines. The main thrust of the adventure will then be to use intuition to discover the underlying reality of the cosmos. Afterwards there will be the possibility of discovering how the universe is really designed and works.

2

Realisations

IT SEEMS PROBABLE that ever since becoming consciously aware, man has looked out on the world around him and questioned his place in it. Two contrasting views could have emerged depending on whether pride or humility held sway.

It is not difficult to see how an overwhelming feeling of superiority, indeed arrogance, could have arisen in those early days. Self-awareness may have given rise to the belief that, as a privileged species amongst others who were different, there was every reason to behave self-importantly. Other animals were obviously eaten because they were looked on as being there for that purpose. The fact that these animals might feed on each other simply because, not being aware, such evaluations were not possible, could easily have been overlooked.

An early fascination with the workings of the universe was probably accompanied by a belief that the whole structure was centred on mankind. In more humble moments individuals would undoubtedly have realised that there are forces in nature that are much mightier than themselves, over which no control was possible. Indeed, whereas in later times mankind could look for the fundamental causes in the very small and in the distant past, at that time there was nowhere to look other than to the world around him.

Unusual natural occurrences in the weather, the heavens and the behaviour of animals were looked on as forewarnings of one sort or another. Floods and droughts and other natural disasters such as earthquakes may well have been seen as retributive. The heavenly bodies up in the sky from where the weather came would naturally have been associated with the causes of these phenomena. In consequence these 'gods and demons' looking down were revered and feared. Archaeological data on past civilisations has indeed exposed places of worship designed with extraordinary ingenuity to sight offended stars at precise times and seek forgiveness. Many religions were established to placate these heavenly bodies as they moved around the sky.

Later, when long and careful observations revealed that these movements were cyclical, the rhythms were noted and recorded. In time these records came to be used to predict the patterns of the seasons and rational explanations for the behaviour of these heavenly bodies, the sun, moon, planets and stars, began to replace the earlier ones based on supernatural forces.

The development and advancement of mathematics and deductive reasoning eventually improved levels of understanding far beyond the needs of farming communities and their harvests. In Ancient Greece knowledge was then pursued for its own sake. In his Academy, Plato gathered around him a number of astronomers, geometers and mathematicians. Aristotle, one of his pupils, wrote extensively on the universe in the fourth century BC. He proposed the view that our planet was not flat, as was popularly believed at that time, but spherical, and placed at the centre of the universe with all the other bodies in the heavens revolving around it in various concentric spheres of their own. This view was maintained and propagated in centres of learning throughout the western world for almost two thousand years.

A major shift occurred during the fifteenth century in the Renaissance. Copernicus had the audacity to proclaim that the movement of the heavenly bodies was better explained by assuming the sun to be stationary and planet Earth in motion. Galileo's support for this view, encouraged by his findings made with the aid of the newly invented telescope, led to a long controversy with the Church that continued to uphold Aristotle's view. The heretical philosopher Giordano Bruno, who proclaimed that the universe was of infinite extent and not even centred on our sun, was promptly burnt at the stake.

Sir Isaac Newton, the renowned English mathematician, scientist and philosopher who was born in the year Galileo died, advanced cosmological thinking through mathematical expression to a greater degree than anyone else. Of central importance to this development was his presentation of an entirely understandable theory of gravitation.

A key aspect of the new mathematical physics developed by Galileo and taken up by Newton was the consideration of how physical systems altered with time. This led to a change in direction in seeking for fundamental causes, away from the eternal present into the past. The German philosopher Immanuel Kant, who was active in the century after Newton, followed this line of enquiry and introduced the notion of a possible beginning and end to the universe. This, predictably, raised the question, 'What did God do before its creation?' In consequence, the idea that the universe was in a steady state continued to prevail. In the early part of the twentieth century Einstein was convinced of this even though his general theory of relativity strongly suggested the contrary. However, it was not until 1929, when a growing body of evidence showed that the universe was expanding, that its 'big bang' beginning was introduced.

Many theoretical physicists now are inclined to believe that the big bang was the beginning of everything. A popular view at present is that it erupted from a microscopic 'nugget' of unbelievable density, the minute size of which would make a grain of sand look colossal. The question as to the origin of this 'tiny bit' is answered in the language of quantum mechanics. It is described as 'a quantum fluctuation'.

This 'fluctuation' is an aspect of the behaviour of matter at a very small scale. There is a limit to how precisely these purposeful structures can be measured and their movements are impossible to anticipate fully. Indeed, their behaviour is unpredictable and the uncertainties appear to be fundamental. They must have profound implications for the way in which the universe can be viewed and these effects become particularly significant in the very early stages of the universe.

A full understanding of these processes needs a theory that combines the laws governing very large objects (general relativity) and the laws governing the very small (quantum mechanics). At some fundamental level atomic behaviour and that of galaxies need to conform; the same laws must apply. This has to be a requirement if the belief that everything in the cosmos originated from a quantum fluctuation is to be upheld. Einstein sought the much-desired unified theory of our universe for three decades. He failed, but a great deal has been learnt since then and much more sophisticated instruments are now available to observe, measure, and analyse behaviour patterns.

A great deal of time and effort has been given to encompass in a single theory all the fundamental forces of Nature. There are only four of these: gravity, the electromagnetic force, the weak force, and the strong force. Gravitation directs its influence on orbiting planets and keeps the feet of Earth's inhabitants firmly planted on the ground. The mass of an object determines how much gravitational force is exerted or felt. The electromagnetic force operates at the atomic level. It powers computers and telephones, and shows its awesome might in thunderstorms. Importantly, its photons, the smallest bundle of the force, make it possible for many animal species, including man, to see the world in which they live. The force causes the countless shapes and textures of the universe miraculously to appear as objects before the eyes. The rainbow colours, which everyone looks forward to seeing when weather conditions allow, also result from these reflected photons. The electromagnetic force and gravity will be much discussed during this cosmic investigation but the strong force and weak force are not well known. This is because their strengths only act at the sub-atomic level. The strong force glues quarks together inside the atom's protons and neutrons, which are crammed inside the nuclei of atoms. The weak force is largely a force for radioactive decay.

Why there are four forces and why they have such different properties remains a puzzle to the cosmological world. Electromagnetic repulsion is immeasurably stronger than gravity but, as this force is in equal amounts of positive and negative charge, it is sensibly contained. Gravity is a force that only attracts: more 'stuff' means more attraction from this fundamentally feeble force. At the sub-atomic level, the strong force is about one hundred thousand times that of the weak force. However, it is only about a hundred times stronger than the non-nuclear electromagnetic force.

Looking back, the desire to connect with the unity underlying our universe has been a strong driving force in mankind for hundreds, if not thousands, of years. The attempts of modern scientists have first and foremost made use of the methods of observation and deductive reasoning, as well as mathematics, and it has led them in a particular direction. However, many doubts remain.

Can the evolution of cosmic behaviour be other than rational?

To answer this question surely demands a more explicit understanding of knowingness.

3
Mankind's Home

THE MILKY WAY, the galaxy in which mankind lives, consists of a few hundred billion stars and is a flattened, disc-shaped system approximately 13,000 light years thick in the middle. This immense aggregate of stars, gas and dust is 98,000 light years across and orbits round its own centre once every 225 million years. In a very few cases non-luminous bodies have been detected orbiting the stars. Our solar system, to which planet Earth belongs, is one example. It consists of the sun and eight orbiting planets of which Jupiter and Saturn are the largest and our home planet is one of the smaller ones. This system is an insignificant blip positioned roughly two thirds of the way out from the centre to the edge of the galaxy and is about five billion years old. It has orbited with the Milky Way ever since its formation and there have been twenty to twenty-two such journeys to date.

A mere hundred years ago mankind thought the Milky Way was the entire universe and believed it to be static and unchanging. It is now thought to be dynamic and constantly changing. It is also recognised that stars have a finite lifetime. They are born, shine for a while and eventually die, not so very different from living creatures. Earth's sun, which is a star of only medium size, is roughly half way through its life.

Man has looked up at the night sky and discerned two types of heavenly bodies since ancient times. Stars, like our own sun, shine brightly because they generate light and heat internally through the process of nuclear fusion. Planets such as our own and our moon are only visible because they reflect the light of stars.

Just after it formed, mankind's home was an airless ball of molten rock. As the rock cooled and a crust was established, an atmosphere began to form from the volcanic gases ejected from within. Later, water vapour fell as rain and formed the seas. Volcanic gases were refined by sunlight into primordial nitrogen that makes up seventy-eight per cent of the air that now supports organisms. Carbon dioxide kept the young planet warm as it gradually became a suitable home for life three and a half billion years ago. The oxygen was all produced by the incessant work of the early living photosynthesizers and levels slowly built up. Respiration, the oxygen-burning process, then provided such an advantage that, two and a half billion years later, multi-cell plants and animals exploded onto the evolutionary scene.

To all intents and purposes our planetary home is now stabilised but, surprisingly, little is known about what it really is. Its diameter is about 12,800 kilometres but the deepest explorations by borehole have reached only about 20 kilometres, less than one third of one per cent of the distance to the centre. Deeper explorations and studies, primarily based on seismic waves, indicate there is a solid inner core. It has a radius of about 1,600 kilometres and is surrounded by a liquid outer core with a thickness of about 1,800 kilometres. This core is probably rich in iron and its temperature is about 5,000° centigrade. An injection of heat provided by radioactive isotopes decaying into stable elements has maintained this level for the last four billion years. This energy supply is slowly depleting and it is projected that in about ten billion years' time Earth will cool down and freeze into a solid lump. However, by this time the sun will also have depleted its own energy source and will no longer be shining on us.

Around the core is a layer known as the mantle that is 3,000 kilometres thick and accounts for sixty-two per cent of Earth's volume. A thin crust covers the entire surface of our planet and its thickness averages out at about 20 kilometres, relatively as thick as the skin on a bowl of cooling custard. It consists of separate sections called plates, six larger ones and twelve smaller ones. All float on the mantle, constantly moving like conveyor belts, driven by the thermonuclear convection currents underneath them. Throughout the life of Earth they have changed the shape of continents, oceans and mountain ranges. In so doing, catastrophic earthquakes and volcanic eruptions have been caused, such as the tsunami disaster of December 2004.

Above its crust, Earth's atmosphere is responsible for the well-being of its organic life and, though this air rises to less than 10 kilometres, it is absolutely crucial in making the planet a suitable home for plants and animals. Driven by a combination of Earth's spinning motion and convection currents powered by the sun's rays, the atmosphere's incessant motion is the cause of our weather and local climates. At times this same movement of air can also become very dangerous, as illustrated by the extreme weather patterns associated with tornadoes.

The ancients firmly believed that Earth, our planetary home, was at the centre of the universe. After Copernicus discovered that our planet revolved around the sun, it took almost two hundred years for this new view to gain acceptance. Similarly, though it was announced to the world in the 1920s that there are other galaxies out in space, it was not until the 1950s that it was no longer possible even to believe that the Milky Way was a giant amongst other lesser galaxies. Its importance in the universe has now been eroded to that of a very average sized spiral and the Hubble space telescope did indeed discover at least a hundred billion others after its most unfortunate lens problem had been overcome in December 1993.

However, it was somewhat earlier that possibly the most profound realisation ever was finally accepted by the cosmological world, and this was that the universe as a whole is not unchanging. It was clearly seen to evolve with time and expand outwards. Einstein told the world this in 1915 with the introduction of his general theory of relativity that, in a series of equations, described the fundamental relationships between space, time and matter. Fortunately for those who are not mathematically inclined, there is a beautiful model to give a simple picture of these mathematical expressions. The universe is represented as a balloon with spots for its galaxies scattered across its surface. As the balloon gets bigger, the skin expands and the spots move further apart. Einstein saw this as space-time being stretched. As the balloon's expansion causes the distance between any two spots to double, the distance between all of them is seen to double. The further a spot is away from another the faster it will seem to be travelling. When the distance doubles the apparent speed doubles.

Like the balloon in the model, the universe has no identifiable centre; its skin is all there is. An expanding universe must of course have had a beginning, a mathematical point in space at a definite time in the past. From studies of the speed of expansion, different ways of computing this 'big bang' all point to a starting point occurring twelve to fifteen billion years ago. Since then the already mentioned Hubble telescope explorations have established the exact time was twelve billion years. Much cosmological endeavour, based on Isaac Newton's laws, Maxwell's calculations and Einstein's two theories of relativity, has scientifically explained much of what happened in the early stages after the initial point of singularity. The problem child of the first one-thousandth of a second still remains and continues to defy explanation.

The general belief is that the big bang beginning of the universe occurred about half a million years before electromagnetic radiation and matter parted. During this time the consensus view is that the temperature measured one thousand billion degrees Celsius and it is suggested that the density of matter was one hundred thousand billion times that of water. Cosmologists are agreed that most matter annihilated itself, everything in the universe today being the one particle in a billion that was left over together with the high-energy photons, a billion in background radiation for every nucleon annihilated. As time progressed, the temperature of this young universe cooled to about a billion degrees, at which time there were eighty-six protons for every fourteen neutrons. Pairing up into stable nuclei such as those of helium then commenced.

For hundreds of thousands of years after helium formed it is believed that the universe expanded quietly. About 500,000 years after the big bang, when the universe had cooled to the temperature of the sun, electrons and nuclei were able to stick together in electrically neutral atoms with little or no interaction with electromagnetic radiation. Billions of photons from the primordial fireball freely streamed through space. Atoms then clumped together to form gas clouds, which eventually collapsed under their own gravity to form galaxies, stars and everything else that is now seen or visualised.

It is perhaps important to record what could prove to be a hiccup in cosmological thinking. It was highlighted during the

Hubble telescope experience mentioned. When the people of the world saw the television releases of the universe, the quality, management and orchestration of the images filmed exceeded the wildest expectations of interested viewers. However, when endeavouring to establish its age, the telescope was directed towards the big bang point from which the expanding universe emerged. Galaxies were then seen where none should have been. The original big bang point is considered to be entirely the wrong place for such structures. To explain why, it is necessary to refer again to the balloon model. Its expanding skin is the universe, which contains all the planetary clusters and stars within countless galaxies that have since expanded outwards. The ever-increasing space inside this skin of activity is considered to be empty of such things, as would be the case when looking towards the centre of a balloon. It is therefore puzzling that observations related to galactic structures seen appear not to have raised any significant questions. It is a matter that will be returned to later in this cosmic adventure.

Einstein said, 'One can organise to apply a discovery already made, but not to make one.' Quite obviously it is impossible to disagree with such a statement.

4

Building Blocks of the Universe

THE FIRST STAGE of this journey was a survey of mankind's home, not just the planet but also the entire universe. Sceptics might say that this probing of the outer regions of space has little bearing on everyday activities, but I doubt this. The attraction of looking up at the night sky and trying to fathom its mystery is irresistible. It is a feeling that should be respected even though my thoughts took me in the opposite direction: from looking outwards and upwards to looking downwards and inwards, to the very small and equally mysterious world of atoms. I was attempting to solve the mystery of how these eternally moving sub-microscopic entities produce the illusion of the solidity of things such as the old oak table that first set me on this journey.

Their importance can be judged by the fact that these atoms construct absolutely everything to be found on this planet, from its iron core to the magma lurking beneath the crust, all living creatures and the water and air that support them. They move around in perpetual motion attracting each other when short distances apart but mutually repelling each other on being squeezed together.

How such widely different things can be reduced to tiny atom structures has been debated interminably. Their mechanistic and teleological basis has long been argued from both the ecclesiastic and atheistic viewpoints. The most simplistic view is that there is no controlling power; the atomic motion is essentially random, ruled only by the laws of chance. However, since the introduction of quantum mechanics it has been agreed that mathematical laws regulate the movements, even if some of these are statistical in nature.

The atom has fascinated mankind for a surprisingly long time. The Greeks gave consideration to the very fundamental question of whether matter could be divided into ever-finer pieces ad infinitum or whether it would eventually be found to consist of indivisible pieces. As early as 450 BC Democritus of Abdera forcefully argued that it would eventually be found that everything consists of small particles. The word atom is derived from the Greek *atomos*, a word that means 'inseparable'. It was actually suggested by this early civilisation that different atoms joined together and, in so doing, possibly changed one substance into another. Plato also describes atoms, but as geometrical entities.

Throughout the nineteenth century, a time when the natural sciences were making rapid progress, scientists made full use of the atomic concept but few were convinced of the reality of these tiny bits. To most they were looked on as being merely a useful tool for understanding the phenomena of physics and chemistry, but not ultimately real. It was not until 1955 that the atom could actually be 'seen'. This was when Erwin W. Mueller of Pennsylvania State University invented the field-ion microscope. This device produced their image at the tip of a needle. Under a magnification that was five-million-fold they were seen as bright little dots.

At the beginning of the twentieth century, just as the reality of atoms was beginning to be accepted, a very remarkable discovery was made. Atoms were found not to be the solid indestructible particles the Greeks imagined but to be made up of parts. What was even more remarkable was that they were found to consist mainly of empty space. The component 'bits', their sub-atomic particles, move around in this space. The obvious comparison to make was with the planets of the solar system moving around the sun. It is now known that there are ninety naturally occurring types of atoms and over twenty man-made ones.

Atoms consist of a central nucleus made up of protons and neutrons around which the electrons orbit continuously. The nucleus is by far the heaviest part of the atom. It consists of a number of positively charged protons, plus neutrons that have no polarity: they don't attract or repel. However, having nearly the same mass as the proton, they are important. They cause the occurrence of different isotopes of an element, slightly different forms of the atom but with the same chemical properties. The calming influence of neutrons still leaves a very volatile nucleus that would inevitably blow itself to bits without the restraints exercised by the strong nuclear force. It will be remembered that this force was briefly mentioned earlier: it glues quarks together inside the atom's protons and neutrons, which are crammed inside the nuclei of atoms. It overwhelms the mutual repulsion of the positively charged protons to the necessary degree, as if they had been constrained by a strong elastic band.

The numbers of electrons and protons are usually equal and this number determines the atom's elemental characteristics. If there is only one of each it is hydrogen, and if there are eighty-two it will be lead. Six electrons and six protons are required for a carbon atom.

The orbiting electrons occupy spherical shells and sub-shells but the designated distances of them from the nucleus are not precisely defined. The accommodation and combination of orbit patterns and their regulatory rules are complex. An extraordinary amount of research has been carried out on these arrangements. Even so it is worth briefly discussing the shell structures to give some feeling of the sophistication of these building blocks of the universe.

As already stated, hydrogen is the simplest atom of all. Its innermost orbiting level can in fact accommodate two electrons, so having only one electron it is only half full. The atom's tendency is to complete this 'shell' and what happens is that two hydrogen atoms pool their single electrons and, by sharing, mutually fill their inner 'K shells'. Hydrogen gas, in consequence, almost always exists in pairs of atoms, as the hydrogen molecule. It takes a great deal of energy to separate this bond. In comparison, helium fills the K-shell with its two electrons and, in consequence, this atom does not easily combine with other similar ones. It exists as a very inert or 'noble' gas.

Lithium has three electrons, two in the K-shell and one in the next shell, the L-shell. Succeeding elements add their electrons to the L-shell until neon, having ten electrons, fills the K and L shells and becomes an inert gas with properties like helium, which had its outer shell, its only shell, filled with two electrons.

Every atom with an unfilled outer shell has an inclination to enter into combination with other atoms in such a manner as to obtain a full outer shell. The lithium atom easily surrenders its one L-shell electron so that its outer shell is the K-shell. On the other hand, fluorine tends to seize an electron to add to the seven of its outer L-shell to complete the eight-electron requirement. Since the nucleus's positive charge of protons does not change, lithium, with one electron subtracted, carries a net positive charge, while fluorine, with one extra electron, carries a net negative charge. The mutual attraction of opposite charges holds the two ions together in a compound named lithium fluoride.

The regulations required for a possible ninety-two orbiting electrons have been extensively scrutinised and it is now known that seven electron shells are used. The innermost one was traditionally labelled K while the others, moving outwards, are L, M, N, etc. The K-shell has a total capacity of two electrons, the L-shell eight, the M-shell eighteen, the N-shell thirty-two and so on. To add to the complexity, shells are divided into sub-shells. In some cases these may overlap. In other words there are deliberate overlaps that are possibly designed to determine where and when exchanges of electrons can be arranged to create desirably workable molecular bonds.

The decisive thing that differentiates elements, as far as their chemical properties are concerned, is the configuration of electrons in their outermost shell. Carbon, with four electrons in this shell, is completely different in its properties from nitrogen with five. In sequences where electrons fill inner sub-shells through interleaving, the properties are less variable. Iron, cobalt and nickel have the same outer shell electron configuration and N-sub-shells with only two electrons. Their internal electronic differences, in an M-sub-shell, are largely masked by their outer electron similarity. That is why, chemically, they are very similar.

It seems important to acknowledge that order and unity are fundamental to atom activity. Each atom cannot be regarded as having a separate existence. All atoms of a type are so identical as to be indistinguishable, even comparing those from different sides of the universe and separated by billions of years. A carbon atom in the first living cell would have been identical to and indistinguishable from the ones now in our bodies.

All the different types of atom are composed from the same three building blocks, namely the electron, proton and neutron, and these common particles are completely indistinguishable in the different atoms. It is meaningless to talk about the electrons of a carbon atom as if they are somehow different from those of an oxygen atom.

Also it is crucial to realise that the chemical properties exhibited by the different atom types are all interlinked. They all belong to a single system represented by the Periodic Table. This shows all the different atom types arranged in order of the increasing number of protons and electrons. In this way they naturally fall into groups. All atoms in the same group show similar behaviour. Thus lithium sodium and potassium give rise to highly reactive metals, and helium, neon and argon form very unreactive gases. The system shows an inherent order. It is not something that evolved, and even if some types of atoms disappeared altogether the system would remain. So it is very interesting to ask where the arrangement came from and how and why it was put in place.

Another remarkable aspect of atoms is that they do not dissipate energy: they do not run down. They exchange energy but only in precisely defined amounts. Their electrons can move to a higher orbiting level but must absorb a quantum of energy in order to do this. When electrons move closer to the nucleus, to a shell of lower energy, they emit a quantum of energy as radiation. At the end of exchanges they return to their original state of rest. To put it another way, atoms do not age during the life of a universe.

Atoms of suitable types latch on to each other to make molecules that are the simplest structural unit to display the characteristic physical and chemical properties of a pure substance. They are the tiniest possible bit of their particular 'stuff'. In living things some of the constituent molecules are quite complex and consist of up to many thousands of atoms. The presence of carbon atoms is particularly significant since these have the capacity to join together to form very stable chains and rings. One of the most complex structures and one of the most important is that of deoxyribonucleic acid, usually referred to as DNA. It is 'nucleic' because it is the main chemical component of the chromosomes that occur in the nucleus of the cell. The acid part is phosphoric acid, which joins with deoxyribose, a cyclic sugar molecule with a structure of a type similar to glucose. The two units alternate in extended chains. Two such chains twist together to form the double helix structure.

Attached to each of the sugar units is one of four bases named guanine, adenine, cytosine and thymine. These point inwards towards the common axis of the two helices. Each pairs up with a complementary partner from the other chain, joined by a special type of bond, just strong enough to hold the molecules together but weak enough to break when the whole structure unravels. The arrangement of bases is neither regular nor random. Their sequence is a molecular code that stores all the information the cells need. Because of the complementary pairing across the double helix the information is effectively held in duplicate.

This means that when a living cell reproduces there is a means by which the information is readily replicated. During cell division the double helix divides. Each free end of the DNA is then able to pick up from the 'chemical soup' the nucleotide with the complementary base and so reconstruct its 'other half'. This produces two identical DNA molecules, one for each of the two new cells. All living cells are built to continue to replicate in this way. Why they do this is one of life's mysteries.

Nearing the end of this stage of the journey it is time to pause and reflect on what has been discovered. Let us recall that the quest was not only for the extent of knowledge but also for the nature of knowledge itself. Different thoughts struck me. Firstly, how different this knowledge of atoms is from the knowledge that is continuously presented through the senses. As I run my finger over a finely woven silk handkerchief, it is extremely difficult to accept the mass of unending activity in the tissue beneath the skin. A piece of wood, or even a sheet of glass in a window, gives no hint whatsoever of the pulsating electrons ceaselessly maintaining the integrity of their elemental atoms. That is the problem when confronted with the world at the level of the atom. It cannot be seen, felt, tasted, heard or smelt. This is what I was trying to tell my friends before this journey within these complex structures began.

In this respect, the air that makes planet Earth able to sustain all its life forms is worth a thought. When its breeze is felt on the face it gives a feeling of uniform comfort to the skin, yet science tells us it consists of separate particles. Each molecule of air is involved in 3·5 billion collisions every second and experiments using smoke particles and a microscope can give indirect evidence for this molecular buffeting, the so-called Brownian motion. And Einstein made a brilliant calculation that showed the movement conformed to mathematical laws.

So what is this illusive world and what makes it so difficult to visualise? Why does it become less definable as knowledge expands? Are these descriptions of atoms 'reality'?

Models have been found to be extremely useful for they focus the mind. It doesn't matter whether they are tangible or just pictures in the head, providing they can be thought around and changed at will. All are invaluable tools of those who make it their business to try to establish what the universe is and how it works. These people desperately look for the right images on which to focus their thoughts. One difficulty is that these aids become more difficult to design as the problems become more complex. The language of mathematics is then extended and can become very abstruse. The situation then arises that the model that was invented to make understanding the world simpler often has the opposite effect.

The physicist's model of atoms has now gone far beyond the semi-classical atoms consisting of electrons orbiting a nucleus of protons and neutrons. They have now been reduced to vibrating quarks or loops. One vibrating loop replaces each electron and three replace each proton and neutron. The strong and the weak nuclear forces then maintain the integrity of these multi-tudinous loops. So the atom, the basic building block of our universe, has been reduced to a complex field of forces. It could be seen as a twanging guitar string except that there are no tangible strings in this redefined world, only twangs. These particles, which are regarded as the basis of substance, have turned out to be insubstantial themselves. They consist of form only, the complexity of which can only be described in the language of mathematics. However, it is possibly helpful to know that it has been claimed that ten million atoms are required side by side to span two dimples in the serration of a postage stamp. It is also said that their tiniest bit is as much smaller than the whole as the whole is smaller than the universe.

This latest thought returns me to the early Greeks. How did they intuitively decide that extremely tiny bits some-how existed to fabricate matter into its cosmic designs? There was no obvious support for such notions; they had none of the analytical tools of modern science. What

plausible reason had this early civilisation for the advancement of such philosophical reasoning?

Intuitive visualisation would seem to be the only explanation. But from where could these 'pictures within' derive?

5

Space and Time

THE OUTER REACHES of the universe and the inner realm of its subatomic particles have now both been visited. Although they are at opposite ends of the scale of magnitude, what they have in common is that the events of both take place in space and time. These two concepts now need to be considered.

It is not easy to visualise the development of a universe after its big bang beginning, but the basic notion somehow seems acceptable, possibly because some of the cataclysmic events that have taken place on our very own planet Earth offer a hint of understanding of the forces involved. The world of atoms, on the other hand, with their orbiting electrons, do not so easily relate to man's sensory perceptions. The ultimate components of these tiny things have been described as strings, force knots, and vibrating loops. Indeed, they are all of these things as they ceaselessly work and adhere to regulations that, as yet, are beyond our comprehension.

Can the understanding of these two, 'the large and the small', be dovetailed into a single unifying theory? The vast difference in size, experienced when measuring structures at the atomic level against those of the universe with all its galaxies, gives an indication of the problems that might be encountered when trying to combine them. The models dealing with each of these regions are already highly complex in themselves; how much more complex would be a model that encompassed both? Yet the basic concepts of space and time which both share seem simple enough. But are they?

Consideration of time and space in modern science begins with Isaac Newton. Even three centuries afterwards no one else's work compares with that of this extraordinary man. One of his main achievements was to apply mathematics to the behaviour of the universe in a way that rationalised its cyclical patterns. He enabled mankind to accept that a single force powered the universe, that everything was tethered to everything else, all stars and planets to all other stars and planets. His gravity did, indeed, reach out across vast distances to exert its hold. Its strength relates to the total mass involved – the sun's grab on Earth being much greater than Earth's grab on the sun. This grab determines their relationship to each other, and particularly the path of Earth's orbit. Newton also showed the world that the attraction depends on two things only, the amount of 'stuff' within each and every structure and the distance between them. 'Stuff' does, of course, mean the number of protons, neutrons and electrons. Mathematically, he was saying that the gravitational force between two bodies is proportional to the product of their masses and inversely proportional to the square of the distance between them. It is a law that has since been used successfully for predicting the planetary motion of Earth and its moon around the sun, as well as for exploring the solar system and studying the movements of stars.

The theory of gravity discovered by Newton is lacking in one very important respect. It offers no insight into what gravity actually is. This extraordinary man was well aware of this fact. He believed that gravity must be caused by a material or immaterial agent acting constantly according to certain laws but, despite the fact that he had developed equations that accurately describe all its effects, he confessed to having to leave the consideration of what gravity actually is to others.

The framework in which Newton worked was one of absolute space and absolute time. He defined them as follows:

> Absolute, true, and mathematical time, of itself, and from its own nature, flows equably without regard to anything external, and by another name is called duration.
>
> Absolute space, in its own nature, without regard to anything external, remains always similar and immovable.

More than a hundred years after Newton's death the physicist Albert Michelson conducted a series of very sensitive light interference measurements in an attempt to detect this movement through absolute space, or 'ether' as it was then called. The context was not now gravity but the search for the medium in which light and other forms of electromagnetic radiation moved. Michelson used an interferometer of his own design to detect and measure directional differences in the speed of light travelling towards and against Earth's rotation. If our planet was moving through stationary ether, a difference that related to Earth's own motion was expected, but despite the most painstaking experimentation the speeds towards and against Earth's motion were shown to be unchanged. The null result called into question the existence of 'ether', Aristotle's 'fifth element'. It also sowed the first seeds of doubt in Newton's concept of gravity, which had been based upon the notion of an absolute unmoving space.

When Einstein's 'special relativity' theory was introduced in the year 1905, Michelson's experiments had already set the scene to change completely the accepted understanding of how the universe works. In Einstein's theory space and time were no longer separate entities: they had become aspects of a four-dimensional space-time continuum in which light played a very special role. Special relativity insists that the speed of light is the same for all frames of reference, namely 299,800 kilometres per second, and nothing else in the universe is allowed to travel as fast. No longer considering space and time as separate absolutes seemed to have some bizarre consequences. In the new language of relativity, objects at rest in space may be considered as moving – moving through time. Two objects at rest must age at the same rate (or speed). If one object on its own moves through space, some of the previous motion through time is diverted and its relative movement through time will be slower. A clock travelling with the moving object will tick less quickly, implying that ageing slows down. Special relativity also demands that light does not age because nothing travels faster; in fact light itself will travel so fast that there is no passage of time, only the eternal 'now'.

The accepted view on the form and appearance of time and space resulting from Einstein's special relativity is that there is no absolute space and time in the way Newton formulated these concepts. They may, however, actually vary in different frames of reference. The absolute now is the speed of light, so, although this goes against man's intuitions, all observers are expected to measure the same speed of light no matter how fast or in whatever direction they move. The implication is that light's relative speed does not change one iota whether it is chased after or run away from. The understanding of space and time has completely changed; no longer are these two measurements thought of as absolute concepts. The theory had implications for Newton's gravity that had been presumed to act instantaneously. The fabric of space-time now causes gravity to react at the speed of light simply because it has been agreed that nothing can travel faster. As a result, Newton's force of gravity had been called into question.

The concept of gravity was tackled directly by Einstein himself in 1915 when he produced his general theory of relativity. The universal attractive force of gravity was replaced by a mutual interaction between matter and space; like space and time in special relativity, these were no longer considered as fundamentally separate. Matter acted on the space adjacent to it and 'curved' it. The apple now fell from the tree not because of a force acting instantaneously between it and the Earth but because of the way the apple and the ground below distorted the space between them.

The balloon model used to describe the expanding universe allows a visualisation of Einstein's gravitational force at work. Stars and planets expanding within its skin create impressions or dimples in the surface to an extent that is dictated by their mass. All movements within the universe must distort this massive trampoline in ways that affect all other movements. Should any two bodies be sufficiently large or close enough together to be forced into a related orbital pattern, it is not difficult to imagine the balloon-like fabric of gravity causing its texture to incline towards each of them in a joint dimple. This route would then be followed until the stars or planets bottom-out and orbit naturally in their individual distortions.

It is fair to say that few claim to relate intuitively to Einstein's theories and have a complete grasp of the abstract mathematics used in the theory. The very significant problem of a unified theory, which was discussed earlier in this chapter, still remains. Even Einstein was not able to resolve this despite devoting most of his later life to it. The gently curving geometrical form of universal space does not relate to the frantic rolling behaviour of the subatomic world; no theory to unify quantum mechanics and general relativity has yet been found.

The explanation given by some cosmologists, such as Stephen Hawking, to a phenomenon that requires reference to both quantum mechanics and gravitation is considered to contain a possible key to a unified theory. It relates to 'black holes' which frequently capture the imaginations of science fiction writers and film directors. The mode of their formation is most easily seen in the context of the life cycle of a star. This begins when a large amount of gas collapses into itself because of gravitational attraction. Its atoms then collide with increasing frequency and ever-greater speed and the gas heats up. Contracting and heating up stops when the pressure equals the gravitational attraction. This state can be maintained for a long time with the expansion due to heat from nuclear reactions balancing the pull of gravity. However, the time must come when a star runs out of fuel. Massive ones take about a hundred million years. Then, as the star cools it contracts until, after a period of struggle, it settles down as a 'white dwarf' with a radius of a few thousand kilometres and a density of thousands of kilograms per cubic centimetre.

There are large numbers of such stars in the universe, as well as others that contract to neutron stars with a radius of only about fifteen kilometres and an unimaginable density of millions of kilograms per cubic centimetre. One possibility is that stars of larger mass explode but otherwise they are unable to avoid a catastrophic collapse as the gravitational field intensifies. It is thought that even light rays are captured by the enormous gravitational pull and are increasingly bent inwards. As the star shrinks to a critical radius light becomes totally imprisoned and a black hole forms.

These awesome holes show gravity at work, pulling on every single particle surrounding it, whether it is a minute speck of dust, a massive star, or a smaller and cooler planet. Its regulatory force is extraordinarily long-range and, if there were no restraints, all matter would be sucked into these accumulations and compressed without limit. It has been suggested that the hole itself would then disappear. Galaxies, star clusters and such exist because their rotational motion counteracts this force. It is a balance that many believe allowed the universe to develop.

A considerable degree of speculation continues to be enjoyed about black holes. Discussions have even been conducted based on what would happen should it be possible to hover just above the event horizon of one for a year and then return home. Many accept that, because of the dramatic time warp caused by the gravitational field's strength, ten thousand years could elapse during that one-year's absence. It is also popularly believed that to achieve such a time warp would necessitate a gravitational compression that would reduce planets similar to Earth to a sphere of less than a centimetre in diameter. It is, of course, understood that organisms do not and cannot live within black holes, so such visualisations have to be unreal. Although completely acceptable in science fiction films, care should be exercised not to let such speculations detract from considering the true nature and purpose of these extraordinary things. Indeed, a great deal of confusion might possibly be avoided if the focal point of cosmic behaviour was not always considered in relation to organic life.

Black holes are unquestionably mysterious and it must be remembered that their number in the Milky Way galaxy alone is considered to be more than a hundred thousand million. This could mean that there are more in the universe than there are stars, and leads one to question their purpose. They often appear after inconceivable compression has affected burnt-out stars. It is also known that they tear up stars that orbit too close. It is a not uncommon belief that the seed of the universe was a quantum fluctuation that appeared out of nothing at all. Exactly the reverse is now seen as possible at the centre of black holes, where it is reasoned that matter can collapse to an infinitesimal point.

Does this give a clue as to how universes work? A quantum fluctuation from nothing certainly does not intuitively relate to their emergence. But it is not difficult to imagine that old worn-out stars and universal debris might collapse under the crushing force of gravity within black hole structures and be subjected to cosmic 'laundering'. But such speculation is obviously premature.

After the big bang beginning of the universe, its matter and energy gradually matured and stabilised into galaxies while travelling outwards, as would the skin of the expanding balloon model mentioned earlier. So should space be seen as nothing? Was there nothing before the universe reached out and will nothing be left behind when it passes on? It is convenient to argue that nothingness is just a word that is used to describe the space between the movements of structures of atoms. Einstein's relativity insists that the envelope of the universe should be seen as a space-time fabric that, while expanding, temporarily adulterates nothingness with the presence of its matter and force fields. Hence the development of the argument that the universe, of which planet Earth is a tiny part, was not created in space-time but that space and time are creations of its presence.

Such descriptions of Einstein's work can only result in extremely negative intuitive assessments. Fundamental problems remain. With extraordinarily practical precision, Newton explained the effects of the force he called gravity, that dominates structures of the universe. Even so, he had acknowledged that others would have to rationalise the 'material or immaterial' agent that caused its behaviour. Einstein's answer was to dismiss 'ether' completely and introduce 'space and time' dynamically. In describing it, the eminent physicist J. Wheeler often said, 'Mass grips space by telling it how to curve; space grips mass by telling it how to move.'

Generally speaking, the cosmological world now believes that motion affects time. The greater the speed, the less the ageing process: time passes more slowly. But mathematical physicists do confess to having difficulty when giving their abstract definitions to time. Since Einstein's theory of relativity the objective scientific use of this concept seems to have become divorced from our subjective notion of simply 'enduring'. It is considered safer to refer to 'measurements by clocks' or 'motion as it affects clocks', and the light or photon clock does illustrate this argument exceptionally well.

These clocks consist of two parallel mirrors, one above the other, with a photon of light bouncing vertically between them. If the mirrors are about six inches apart, one facing up and one above facing down, it takes a photon a billionth of a second to complete a round trip between the mirrors while the clock is stationary. If another identical clock slides past the stationary one at a constant velocity, its photon must travel diagonally as well as up and down to maintain its position between the mirrors. The constancy of the speed of light will ensure that the speed of the photons of light does not change, so the moving clock's ticking must slow down in relation to that of the stationary one. The recorded time of light clocks on the move must therefore be less than for stationary ones.

It would be very convenient to say that it is obvious why the photon clock argument applies and, therefore, reasonable for photon clocks but not for other things in life. However, this is not accepted for it is seriously claimed that all clocks would react identically, as would organic life. The differences in relative speeds on Earth are, of course, so small that any distortions in the passage of time are not measurable. Nevertheless, if it were possible for a traveller to pass a stationary observer at a speed that was three quarters that of light, it is forcefully argued that his moving clockwork watch would tick at about two thirds of the ticking rate of the stationary one. It is also claimed that the traveller would age at a rate that relates to his own watch.

Einstein's theory of relativity goes further. It insists that measurements are only meaningful in relation to a defined frame of reference such as Earth, the sun, or man. Space and time measurements then become 'relative' and another identical planet passing Earth at a speed of 225,000 kilometres per second would appear foreshortened by fifty per cent in the direction it travels. It would be elliptical in shape and twice the mass of Earth. But an observer on that planet would consider himself to be stationary and Earth would be moving past him at 225,000 kilometres per second. Earth would appear elliptical in shape with a mass that was twice that of the observer's planet.

A tempting response to the many radical explanations of space-time might be to ask for a definitive answer as to which planet would actually be foreshortened. Unrelated? Unreasonable? It possibly depends on whether an intuitive investigator or a mathematical one answered the question. Again, this scientific notion of passing time is very different to the innate subjective experience of consciously aware beings.

Not surprisingly there are other 'space-time paradoxes'. Two spaceships pass and both consider the other to be moving at

225,000 kilometres per second towards them. When returning after one hour, observers on each ship would expect the clock on the other ship to have slowed down by half an hour in relation to their own. This is of course impossible because both clocks cannot be slower. So what would really happen? Apparently nothing: ships flashing by each other at such speeds would separate forever and the clocks could not therefore be measured. But the question has been voiced, 'Suppose they could come together?' Somewhat unsatisfactorily it is suggested that either of the ship's clocks could then be slow, depending on which frame of reference is chosen.

Relativity theory insists that there is an ultimate speed, which is that of light. However, in quantum theory it now seems to be generally accepted that particles can and do intercommunicate instantaneously, an anomaly that has to some degree been reconciled by agreeing that the reaction must be before news of the event arrives! In terms of relativity theory, it is also true that Newton's instantaneous gravity is unacceptable.

Careful examination strongly suggests that a number of obstacles exist to a unified theory of everything in the astrophysical world. In consequence, this survey of space and time has raised some important questions about the origin of knowingness. Although the big bang theory is fascinating it always presses the question, 'What was there before it?' Man has the capacity to contemplate time, but does this not suggest that there is something within him that is itself not subject to time but free from it? The same applies to the laws of nature. Newton's laws, including his law of gravitation, describe how mechanical systems change with time but the laws themselves are regarded as unchanging. Does that make them eternal? If so, did they exist before the big bang? The same has been said of atoms and molecules and the system of elements presented by the Periodic Table. These also are regarded as not changing with time. Are these also eternal so that they transcend time? Finally there is our own experience of simply being present, which underlies any subjective notion of passing time. How is this accounted for?

One is left with the suspicion that trying to find a unified theory of everything by combining relativity theory and quantum mechanics in some mathematical-type theory is not going to work. Even if it did there is a suspicion that it would be so difficult to comprehend that the answer would not be of much help to ordinary people. However, it does seem that the universe is a unity because it all came from the same source. Given that organic life is made up entirely of this substance of the universe, perhaps an alternative approach is to seek to find the unifying principle much closer to home.

6

Life

EXPLANATIONS OF the behaviour and management of the universe that unify all its various aspects are extraordinarily difficult to visualise. The tiny atom, massive galaxies, space-time, life's origin, thought, human behaviour and sentience are each lifetime studies in their own right. The world can so easily be looked on as ordinary, something to be accepted as it is seen, heard or felt with mankind's senses, which are themselves totally beyond cognition. So much is taken for granted; yet cosmic reality must be addressed. One approach to solving this puzzle has to be an improved understanding of organic life, for only then will it be possible to visualise the true characteristics of cosmic structures and their force fields. However, an entirely different approach to the one that relies only on the outcome of experimentation and deductive reasoning might be required before definitive explanations become possible. Life is much closer to us than both the world of atoms and the galaxies of our universe; we have a much more direct experience of it and so we should think carefully before dismissing our own intuitions.

We should be wary of putting the cart before the horse and have regard to the nature of knowledge itself rather than just its content. This is particularly appropriate when the object of study is life itself. Three hundred years of purely experimental-

THE CELL

Organic life consists solely of individual cells, and mankind is a multi-billion conglomerate. All contain liquid cytoplasm surrounded by a membrane that regulates the passage of useful substances into the cell and rids it of waste. The complex tasks of an organism's life are performed by its cells, and their chromosomes hold the instructions for protein synthesis. They are passed into the cytoplasm and decoded in ribosome particles.

The DNA of each and every cell carries all our hereditary characteristics.

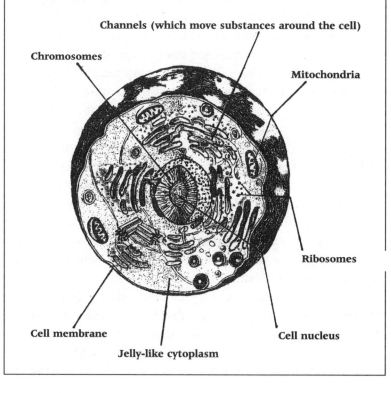

Channels (which move substances around the cell)

Chromosomes

Mitochondria

Ribosomes

Cell membrane

Cell nucleus

Jelly-like cytoplasm

mathematical investigations into nature is far too long for so little return. The suggestion that a change of strategy is necessary must be taken seriously, and in particular intuitions have to be given greater respect. Despite the extraordinary amount of accumulated knowledge and related technologies, it has to be admitted that very little that is really fundamental has yet been found out about our planet's relationship with its universe and the cosmic whole.

Some early impressions during these travels suggested that many of the most inexplicable and bewildering questions relate to organic life. Why did it evolve? What was the cause? How can it replicate and respond? Science has made considerable progress in tackling these questions and this will now be reviewed.

One of the secrets of life which has been very thoroughly investigated is the way all living matter is made up of cells, the basic independent units of life. Just as everything in the universe is built from individual atoms, every living animal and plant is made up of cells. The simplest creatures, organisms of the kingdom Protista that includes bacteria, consist of a single cell, while the bodies of human beings contain several trillions. Clearly visible using a conventional microscope, they are of the order of a tenth of a millimetre in diameter. Those of plants are surrounded by a distinct cell wall made of cellulose, enclosing the cellular material known as protoplasm. Those of animals are usually smaller, with less distinct boundaries, and, whereas plant cells usually have a large fluid-filled cavity, animal cells are almost full of protoplasm.

Possibly the greatest mystery of life is how it began. From where did it arise? Even the simplest of living things are immensely complex, particularly their replicating machinery. There is a view that life appeared spontaneously, but on a rational evaluation this is not very plausible. Poetic descriptions have been voiced suggesting that Earth's primeval surface was washed into oceans to form an ever-richer soup of chemicals, and time was the only additional ingredient required to produce the first self-replicating molecule out of this primordial broth. However, mathematical calculations have shown that the enormous number of chemical combinations that would be needed make the probability of such an event happening to be less than flipping heads on a coin six million times in a row. Surely, just random organic molecular shuffling is too improbable a way for life to have emerged.

It should be acknowledged that the original assembly of organic life's DNA on Earth could not have been inherited. Ancestors are needed before genetic inheritance can be accommodated, at which time Darwinian-type laws can take their rightful place. Acceptance of this fact surely demands a consideration of whence intellect derives – intellect that is capable of visualising genes and the information they store. To suggest that it can also be a chance happening does not seem to be convincingly arguable. That it occurred within a universe that was 'a unique event' is poetically pleasing, but its true origin is as elusive as the emergence of the first organism.

A marvel of organic life is the way it reproduces itself. How the union of a tiny male and female cell results in an integrated organism, particularly one with advanced sight, hearing and mental facilities, also needs a rational explanation. Furthermore, it is extremely difficult to envisage how the natural evolution of embryos from single fertilised female cells 'just happens by itself.'

Billions of identical cells are directed to specific locations within the embryo over a reasonably short period of time. They transform themselves into selective entities suitably designed for their precise position in an eye, a neuron of the brain, a minute part of a skeleton, or some other equally important organ or part. They interconnect in a group formation so that they live, are nourished by and serve their functioning part in whatever way they have been instructed. The organ or part then carries out its own independent responsibilities in a complete organism and, within this factory of billions of working cells, the design is finely tuned to suit the purpose exactly. The functions continue throughout the life of complex organisms that can be either consciously aware humans or less advanced animals or plants.

How can a cell know that it has to become an artery cell, a blood cell, or part of the brain? What directs these identically coded living units to such widely diverse locations with such precision? Even more difficult to attempt to visualise is how mutilations cause embryos to rearrange their pattern of growth or assembly. The hydra, a simple freshwater coelenterate consisting of just a trunk and tentacles, is remarkable inasmuch as, if it is cut into pieces, each part can grow back into a whole organism. Similarly, some reptiles such as salamanders are able to regenerate limbs that have been lost.

It is now accepted that the final form of biological organisms is determined with great accuracy by the encoded directions of their DNA. This genetic intelligence of every cell exercises a co-ordinating influence over billions of cells. All parts have to be sufficiently informed to connect to the complete life form so that it works and responds as a complex whole. The knowledge applies not only to the form of the creature but also to its behaviour. There are such mysteries as the fantastic feats of navigation performed by birds. They travel thousands of miles with pinpoint precision. That the knowledge required is stored genetically and translated by systems of molecules into such behavioural skills is quite extraordinary. A spider's intricate web, built with beautiful precision and effectiveness for its purpose, is achieved seemingly without purposeful intelligence. This instructive or innate behaviour is born into the spider's nerve wiring systems, like pre-set switches.

The process of morphogenesis cannot be explained just in terms of the form that develops from the egg or seed. One of the difficulties is trying to understand why, when part of the developing structure is removed, the system evolves more or less normally. As long ago as 1890 a classical example of this complexity was shown during experimental research on sea urchins. At the two-cell stage, one of the cells was killed. The embryo continued to grow, but not as a half sea urchin. The final form was a small complete sea urchin, as was the case when two, and then three, cells were removed at the four-cell level. Later, the fusion of two young sea-urchin embryos surprisingly produced one giant sea urchin.

This behaviour suggests that the DNA acts not just as a passive store of information but that it is able to give identifiable individual instructions to progressive groups of cell divisions during the process of embryo growth. It would all be a bit like knitting a sock. With one stitch dropped when casting on, the sock must end up complete but smaller in size. However, even if the 'dropping one stitch' view were a possible answer to this aspect of morphogenesis, it is a very small step in the direction of explaining the complexities of living organisms.

No one can question that it is possible for computers to solve complex problems at far greater speeds than humans. Genome intelligence, with its ability to deal with the initial development, growth and maintenance of living organisms has many similar characteristics.

The molecular structure of DNA, the molecule that holds the genetic information, was described earlier, but this is only part of the story. It will be useful now to look further at the way it stores its information and how it communicates with the cell.

Chromatin threads were actually discovered in 1888 and they are colourless and quite difficult to examine against their common background unless they are stained. In all human cells there are twenty-three pairs. They contain DNA sections called genes that carry all the information about how to build the proteins that are the main working components of the cell. The DNA is coiled in a very compact way. Placed end-to-end chromosomes would cover a distance of two tenths of a millimetre but if the stored DNA were stretched in a straight line it would cover a distance of almost two metres. In the cell it is all packed into a length of only one ten-thousandth of its stretched-out length, yet despite this its machinery can somehow find the bit that is wanted to make a particular protein. It then unwinds the section of the chromosome that is required, copies the DNA message on to a messenger molecule called RNA, and winds everything back up again in the space of a few minutes. This performance becomes more extraordinary when consideration is given to the vast numbers of genes spread out along the twenty-three chromosome threads.

DNA molecules each contain millions of atoms. As described earlier, they form a double helix structure, like two intertwining spiral staircases. The backbone consists of an alternating sequence of phosphate and sugar molecules with four different bases attached to the sugars and forming the steps of the staircase. The initials of their names, A, G, T and C, usually represent these four bases that constitute the letters of the genetic code. They are designed so that the pairs A-T and G-C loosely link across the two strands of the double helix. Thus, these complementary base pairs cling together at each step of the double helix staircase. When replication of the cell takes place, even though the double helix divides and the two halves separate, each still contains all the information to construct the organism.

The relationship between DNA and proteins is such that the four-letter DNA code is the basis of a twenty-word protein language. Each word corresponds to one amino acid, the basic unit of protein chains. It is universal and the same language is used in all cells whether they are of bacterium, plant or animal. There is now evidence to suggest that, of all the different possible codes, the one actually used is among the most efficient that there can be. The question of how good the DNA code is compared to other theoretically possible variations was tackled by some molecular biologists. They used computer methods to produce at random different variants of the code and then to test their relative efficiency. A million variants were produced and, as might be expected, a graph of the distribution of their efficiency produced a bell-shaped curve with a large number of codes producing an average efficiency, tailing off on both sides to give a few very good and poor ones. Where did the natural code come? The workers were surprised to find it was nearly off the scale; it was just about the most efficient of all the possibilities, literally one in a million. It is very difficult to explain this result in terms of the operation of the law of chance alone.

The process of synthesising proteins from the amino acid building blocks available in the cell is quite intricate. It is an assembly line operation that utilises enzymes, functioning as chemical catalysts, to cement the new molecular bonds in the growing protein chain. The factory process directs the placement of amino acids in a specific order in a molecule that could contain many hundreds of units. For each position it must choose the correct amino acid from some twenty different ones available in the cell. If there were twenty corresponding units in the DNA molecule, the code would be much simpler, but it will be remembered that DNA has only four nucleotide building blocks. If used in pairs there would be sixteen combinations. This would not be quite sufficient, so they are used in threes to give sixty-four codons or 'word' combinations.

The question of which codon corresponds to which amino acid was not tackled until 1961 when biochemists were able to work out the nucleotide triplet that corresponded to each particular amino acid. It was found that the coding was degenerate in that a number of codons could represent one particular amino acid.

And how does the information reach the sites of enzyme synthesis in cells? The answer to this second question is 'deviously'. It is a result of the unwinding process of the two strands of coiled DNA. One of them replicates its structure on a nucleotide of an RNA molecule, which can then leave the nucleus of the cell and enter the site of enzyme manufacture. This site is a tiny particle about two millionths of a centimetre in diameter, called a ribosome. In mammalian cells there are as many as one hundred and fifty of these sites of protein synthesis.

Cells manufacture protein with an ultimate capacity to combine as many as fifteen million amino acids a minute. There are some that scarcely manufacture any, yet all cells in a given organism have the same genetic organisation. An individual cell may synthesise a protein rapidly at one time and slowly later, while a cell with an unchanging gene pattern can produce quite different combinations at different times. The flow of information is not only from gene to enzyme. A prime example is the formation of an enzyme that catalyses the reactions inter-converting the amino acids threonine and isoleucine. There is a careful regulation, a feedback mechanism, so that, as the isoleucine concentration goes up, less is formed. It serves to activate a suppresser, which begins to shut down the very gene that produces the particular enzyme that catalyses the reaction. Alternatively, if the concentration goes down, the gene is unblocked and more isoleucine is formed.

The chemical machinery of all animal cells is quite remark-able. Their genes, repressors and enzymes are complex to the extreme, and all interrelate interminably. In consequence, it is very difficult to argue that the design of a chemical system of such sophistication could ever have resulted from the accepted processes of evolution. Intuitively, it has to be seen as quite impossible. Complex atoms that work ceaselessly and conform to regulations that are beyond mankind's understanding are the building blocks of their organic proteins and the closely related 'genetic blueprint' of each cell's genome. All are indispensable to the objectives of the known universe, and it becomes more and more difficult not to conclude that there is awareness and related intention in this.

It is now quite well accepted that gene data shapes all organic life's behaviour patterns below the level of consciousness. The introduction of automated techniques and the availability of massive computing power have produced a very rapid increase in the information available from the related area of protein structure. There is an ever-growing catalogue of the structures of the multifarious body proteins that is helping to reveal in ever-greater detail the complexity of the metabolic and signalling patterns of the cell.

In some respects proteins are even more complex and subtle than genes. They are fickle; they can come and go with the vagaries of their organism's behaviour, growth and repair demands. Also there is no straightforward one-to-one relationship between genes and proteins: it is not as simple as that. Genes have the potential to activate hundreds of protein-supply or signalling activities, the total numbers of which are in the hundreds of thousands. Other body chemicals may also modify these demands. Their lawful complexity has remarkable similarities to atom activity and, in consequence, it is not difficult to argue that both behave purposefully to patterns that are unknown to us.

Complexity is so obviously apparent in genetic knowingness that no apologies need to be made for implying the influence of 'intelligence'. It cannot be denied that individual cells 'intercommunicate' within their multi-cell structures, and they do so from programmed gene data that is, of course, inherited. However, an understanding of gene and protein complexity is possibly not as important to this exploration as visualising a model that embraces their common idiosyncrasies, many of which seem also to be shared with atoms.

What was most striking at this stage of the investigation was the realisation that the building blocks of life, the bits that fabricate duplicate living cells, are not pre-instructed. They are not even organic. They are atoms. It is surprising that these smallest of functioning units did produce an original cell 'once upon a time': that surely has to be seen as extraordinary in the extreme.

The unpredictability of the atom has caused a great deal of heartache in the scientific world. Mathematical models for dealing with this have been developed but they are of considerable complexity. Their indeterminacy has never been fully accepted.

Most famously, it was Einstein who voiced his reservations of the probability-based interpretation of quantum mechanics with the famous declaration, 'God does not play dice.' He found it difficult to reconcile this interpretation of the behaviour of atoms with his intuitive appreciation of the lawfulness of the universe. A degree of uncertainty within them might be essential to the working of the universe. Its continuity, its planets, its galaxies, and the creation and development of living creatures, possibly require their 'randomness'.

Whilst reflecting on this subject, the view emerged that perhaps living creatures, constructed from these building blocks of identical atom designs, actually need uncertainty beyond that resulting from natural selection and mating choice. It could be necessary to ensure continuity of evolution in the ever-changing conditions that have to be faced. At this stage of the journey I just did not know, and my field notes recorded the fact. Unpredictability, whether illusory or not, is a type of motive: it encourages incentive and enterprise. It also supports the concept of choice as something voluntary. This is in accord with a belief in mankind's freedom – that we have the ability to choose and make firm decisions that can be acted upon. But perhaps a more fundamental reason for the 'fickleness' of the atom will emerge.

This study of the cosmos has revealed persistent similarities between the many evolutionary processes that dominate planet Earth and the substructures of the universe. There are different levels of organisation. Each is built from the components of a more basic level but has meaning and purpose of its own. Furthermore, this meaning and purpose peculiar to each level cannot be deduced from the component parts. At each level the new whole is complete and is greater than the sum of its component parts. This extends to all but the first moments of the big bang beginning. The so-called 'inflationary theory' that has been propounded to explain these moments suggests that for less than a billionth of a trillionth of a second an explosive, accelerated expansion took place. It is only later, when the universe was about a thousandth of its present size, that differentiation started to become obvious. At this stage it slowly progressed from a sizzling, incandescent fog into particles. These tiniest of bits then gradually combined to make the universe as it is now seen by conscious organic life.

Atoms, with their perpetually orbiting electrons, are generally accepted as being the first self-reliant structures of a universe. The simplest ones had only one orbiting electron but later ones had up to a hundred. These little satellites provided atoms with an inbuilt interleaving facility that enabled complex molecules to be produced. Planets, solar systems and galaxies gradually formed and expanded outwards, but always under gravitation's controlling grip. Although the scale difference between an atom and a universe tends to obscure the fact, the similarities between these sequential patterns are recognisable.

Other common factors become obvious when thought is given to the first emergence of life in the form of the single replicating cell. In terms of complexity this organic marvel makes the atom appear simple, but it must be remembered that every cell is built from atoms and nothing else. There are apparently untold millions of them all doing their own thing, but the outcome is a living creature capable of sustaining its complex organisation and reproducing itself. Consideration has then to be given to the fact that each and every organism, be it plant or animal, carnivorous or herbivorous, derives from one cell only. Just as atoms unquestionably fashion galaxies and in turn galaxies constitute universes, so cells are the building blocks of plants and animals that in turn live their own 'community life' in cities and pastures.

The animal brain and its nervous system, as well as the arterial circulatory designs of the highways of the heart, invoke a comparison with the infrastructures that mankind has consciously and purposefully designed into its cities. When these in turn are compared with the complex structures found in bee-hives, anthills and the nests of termites, it starts to become obvious that complementary patterns prevail throughout the various levels of nature's collective functions: not random patterns but pathways of organisation.

Seeds, the reproductive 'germ' of plant life, are ingeniously designed with exquisite airfoils and other airborne aids to carry nature's 'reproductive eggs' advantageously to pastures new. Those of other plants are cunningly enfolded in tempting fruit, which is eaten and carried by animals in their gut to distant barren areas. Germination is then assured when the seeds are dropped, encapsulated in the fertile dung of the carrier. These behaviour patterns of the life forms of nature all enjoy the common design inclinations that would be expected from a unitary intelligence. Charles Darwin named it 'development through natural selection' but he was concerned with only animal and plant life, just one aspect of the cosmic whole. The phrase is entirely inappropriate when considering the intelligence in the system of atoms or of a galaxy with its stars and black holes.

The following observation relates to a time when a model was badly needed, something to illustrate simply and clearly a thought that, until that time, had been only partly formed. A jumbo jet flying serenely overhead eventually solved the problem. It symbolised so much.

The jet's compact elegance and ability to perform in a way that can only be described as dignified, somehow failed to conceal its complexity. The feelings that arose when gazing at its unhurried but purposeful progress across the sky were very similar to those engendered when giving consideration to the living organic cells of species. Having been accepted as the smallest structures capable of independent life, they had been considered in considerable detail. Their complexity was obviously much greater than that of the jumbo, for not only does each one have more bits and pieces than the whole of that aircraft, but cells are capable of supporting themselves and jumbos simply cannot. Cells also contribute to the functional needs of complex animals, of which they are an infinitesimally minute part. The jumbo jet has no such obligations, though it does support many and varied passenger needs.

Despite the great respect that exists for these aircraft, it had to be admitted that their electronic communications systems cannot be compared to the knowingness of the genomes of each and every organic cell, with their extraordinary encoded text. There is then the indisputable fact that the jumbo cannot even reproduce itself and, should it be possible to group billions of them together, they could never be persuaded to behave as a single integrated unit.

Obviously, such intuitions are not as sharp and final as these words suggest. Intuitions don't behave like that. The 'pictures within' have to be unravelled and digested. Supposedly, it is what is called 'thinking'. But when all this has occurred, tormented thoughts relating to the origin of life can somehow disappear.

Both the cell and the jumbo are composed of atoms and the aeroplane is infinitely less sophisticated than the

organic cell. It surely cannot ever be suggested that a sufficiency of time would result in a jumbo jet just 'emerging' somewhere, as is so often believed of the living cell. The jumbo's evolution needed the combined, consciously aware intellect of the human race. It was caused to arrive by mankind's imaginative efforts. So why is it widely believed that a sophisticated self-reproducing, self-supporting cell 'just arrived'?

The behaviour of flying birds and insects results from genetically co-ordinated aerofoil shape changes to their wings and the whole body. Mankind's aeroplanes control direction and lift mechanically but all are powered organically. Birds eat and digest other organic life to obtain their energy and flying machines obtain theirs from organic matter that has decayed below the Earth's surface for millions of years.

7

Awareness

WHEN HE LOOKS at himself, consciously aware man has to accept that he has a mind as well as a body. Any comprehensive account of the working of the cosmos must include this aspect of his existence; it cannot be ignored. The body is obviously there. It can be touched and seen. It relates to the physical world. When considering the mind, however, endeavouring to understand what is under the microscope can be unnerving. The mind cannot be seen in the same way as the body; it is a 'feeling within' that sometimes compels bodily responses.

There are the unconscious, instructive behaviour patterns, such as those concerned with migration and the complexities of navigation that are inborn within birds. The unimaginably intricate programmes for such behaviour patterns are encoded into the DNA, as is the instructive data for such diversely different activities as metabolism, body regulation and repair systems, and the embryological development of organisms. But DNA cannot account for all behaviour. There are the deliberate decisions related to the outer world that require individual volition and cannot be genetically pre-programmed. In humans there is knowledge gained through experience, example, talking and reading. This enlightenment has to be available in such a form that it can be evaluated and acted on.

Mankind fully understands that most body function activity is governed and co-ordinated without conscious direction. It is also accepted that there is little understanding of what is continually happening. There is awareness when a need exists to eat and evacuate the bowels or bladder, but we rarely consider the complicated digestive processes and distribution arrangements executed every moment of every day. It is all genetically motivated and these activities are co-ordinated according to genome pre-programmed instructions passed on to living creatures at conception. It is an extraordinarily complex multi-cell operation in which each and every cell within an organism plays its part accommodatingly. All have identical DNA blueprints, their chromosomal tapes of gene information. Yet, as already discussed, every cell is selectively instructed to react to information that allows for a collective performance. The total organism then lives as a conglomerate whole, intellectually and mechanistically. It is an operation of mind-bending complexity and some aspects of it need to be surveyed.

Animal organisms made up of more than one cell must have a system that co-ordinates their responses and their 'bits'. All need to intercommunicate. Coelenterates, the simple invertebrates such as jellyfish, have the beginnings of such a nervous system. Stimuli such as temperature, light, water, movement or the touch of some object create nerve impulses that jump tiny gaps along rudimentary nerve fibres. The adult jellyfish drifts in the water with limited control over its movements. It has muscles that allow it to contract its belly, reducing the space under it, forcing water through the opening. This pulsating rhythm throughout the whole organism causes it to swallow food or move away from an intrusive stimulus. In better-developed systems nerve cells exchange thousands of such impulses with their neighbours, sending specific signals to selected areas.

Special groups of cells which comprised simple sense organs originally evolved in flatworms, giving them the advantage of not needing to send impulses throughout the whole body. From this development emerged the first central nerve cord, which became particularly well formed in the region of the head. It was indeed a brain of very simple structure. More complex life evolved as sense organs and nerve cords developed. The knot of nerve cells at the crossroads in the head became more complex as nerve fibres developed and were able to carry impulses in the region of 500 kilometres per hour instead of one fifth of a kilometre. Later, in vertebrates, the cord was protected within a bony spinal column that also served as a girder from which to extend rib bones to protect delicate organs, as well as the long bones that formed limbs. Chordate brains then improved the three structures that had already existed in the most primitive vertebrates. What were originally mere swellings of nerve tissue later became the forebrain, midbrain and hindbrain. Strangely, it seems that the later development was anticipated in the earlier structures. The first of these, known as the cerebrum, is generally accepted as being responsible for acts of thought and will in the most highly developed creatures. As the brain grew larger, it bent into folds to accommodate its complexity.

The brain of the human species has tripled in size in the last three million years, a very fast increase in evolutionary terms. It is popularly thought that this could be due to man's walking erect, which lifted the eyes to new horizons, enabling more information to be delivered to the brain. Freeing the forelimbs to explore and feel the environment gave further help. A flood of sense perceptions were then released as a result of long-distance sight and short-distance touch, which gave man an inevitable advantage over other species. In consequence, these larger brains were favoured by evolution. Mankind's brain is about one fiftieth of the total body weight. Each gram of its weight is in charge of fifty grams of body. The chimpanzee's is about a hundred-and-fiftieth of its total weight and a gorilla's approximately one five-hundredth. In contrast, an elephant's brain is only one ten-thousandth of its weight.

Though humans have large brains in absolute terms, they do have a rival in cerebral development in the dolphin. Similar in weight to a human, it actually has a larger brain that is also more extensively convoluted. It is possible that this brain is more concerned with lower order functions, though experiments suggest that it could be comparable to that of mankind in some higher order functions. These graceful creatures certainly have a complicated speech pattern, and inter-species communication at some future date cannot be ruled out. It can be argued that they lost their opportunity to translate their intelligence into environmental control when they readapted to sea life. For a dolphin the use of fire was never a possibility, whereas mankind's discovery of its use was one of the factors that first marked it out from all other creatures. It is also obvious that the super-streamlining of dolphins ruled out the evolution of anything resembling arms and hands with which to investigate and manipulate.

This survey of the nervous system and brain provides some indication of how co-ordination takes place in living creatures, yet one keeps coming back to the inexorable fact that continually pulsating atoms and nothing else are the basic building blocks of all these sophisticated cellular behaviour patterns. It remains extraordinarily difficult to envision how these tiniest of bits interrelate in living systems. It is even more difficult to envisage how their movements relate to thoughts, sight, memories, hopes and fears. Most difficult of all is trying reconcile our subjective experience of consciousness with these atom movements.

The extraordinary development of complex atom structures over three and a half billion years surely has to be a factor in the arousal of mankind's consciousness. But what exactly does consciousness mean? What is the essence of this distinctive feature that makes mankind stand apart from other creatures? Birds, that are known not to be consciously aware, had flown and navigated globally with great precision before mankind achieved his advanced intellect, and all the other extraordinarily intelligent forms of life seem to have managed without it. So what is its special purpose?

Perhaps, in beginning to consider this question, there needs to be a change of direction from outward to inward, from regarding life as an objective process to considering its subjective aspect. A good place to begin is with the senses. A little reflection shows that there is a major difference between the sense processes and objects 'out there'. Our senses are part of our subjective experience. The focus here will be on seeing, the sense that probably takes our attention the most.

There is a general consensus about what man sees. Problems begin when trying to decide what seeing really is and how it is accomplished. It is now considered naïve to regard the eyes as looking out at a world that is colourful, detailed and directionally explicit. This is rather like experiencing the sun rising in the east, when it is really Earth's rotation that gives the sense of movement.

The scientifically minded person queries this interpretation of sight by suggesting that the very dark world that is 'out there' is not what is seen. The world that is 'seen' is regarded as consisting of 'images within' that are created by electromagnetic waves reflecting off objects such as trees and houses, and even the planet's atmosphere. It is mainly the arrangement of electrons within molecules that determine the extent to which the textures of the universe can absorb them. Those that cannot be absorbed are reflected into the eyes, where they are converted to shapes and colours that are mysteriously conveyed to this 'inner being'. Further information about the form and position of the object seen is gained from extraordinary eye orientation to skull location calculations.

These shape-colour sensations are somehow felt within and allow man poetically to describe the world as beautiful, awesome or terrifying. He can also pick his way through its dangers and interests, guided by the projected 'pictures within'. His experiences can be written about so that others are able to use the process of sight intelligently to read what has been described. This 'seeing' phenomenon appears even more extraordinary when it is realised that many different people are able to build similar pictures in their heads when reading what an author has written about his own imagined experiences.

It is of course our consciousness that makes this possible. Vast numbers of unaware animals are sighted; they hear the world around them and experience the senses of touch, taste and smell – but, not being aware, they cannot appreciate the music of great composers, whose language many believe to be that of God to man. Neither can they experience the sight of the historic pyramids, St Peter's Church in Rome or a Raphael painting, or be invigorated by an angry sea, a thunderstorm or the feel of the sun, wind and rain. These are the sounds, the sights and the sensations to which our consciousness responds. It gives rise to feelings within us that convey the essential fact that we exist.

So how can this knowledge that we have of the cosmos be advanced? Persistent inner feelings insist that the orientations of mankind's thoughts are hopelessly out of focus. Is the obvious somehow being overlooked?

So many of the qualities within us that determine identity and purpose are at present unthinkingly accepted. This is true whether they are attitudes, opinions or sensations. Should not this critical realisation of one's own identity, this cognisance of self, be more heedfully appreciated?

This résumé argues forcefully and repeatedly that 'thoughts from within' are important. Consciousness, the quality or state of being aware, is distinct from sensory perception or thought. This awareness of self allows existence to be both appreciated and feared.

The progression of the universe to its present state of fulfilment is possibly quite well understood at the galactic level. A considerable degree of poetic licence may have been exercised in the attempts to extrapolate back to its big bang beginning, but new ideas and theories continue to germinate. However, perhaps we now need to go beyond ideas that are expressed mathematically, and cautiously give some real weight to intuitions.

Consider the fact that man is created from 'cosmic stuff'. It contrived his genetic ancestry three and a half billion years ago in a universe that had taken more than twelve billion years to arrange. It also orchestrated his consciousness, possibly more recently than sixty thousand years ago.

An important consideration has to be whether morphogenesis, the coming into being of specific forms of living organisms, was always predisposed towards *Homo sapiens* and the emergence of self-awareness. It would certainly be difficult not to accept that aware organic species are an intention of cosmic behaviour. Only atoms existed in the early life of the expanding universe and these were disciplined within its space by electromagnetic waves and gravity, a cohesive force of attraction. The replicating organic cell later gave planet Earth more complex structures, and mankind's consciousness certainly guaranteed an appreciation of this.

Mankind's evolved brain is able to respond to a million billion units of information that relate in some way to a hundred billion neurons. This level of complexity is possibly necessary for an appreciation of consciousness. But consider the earlier suggestion that the dolphin's brain is possibly more developed than man's. These lovely creatures are undoubtedly aware, if brain complexity determines awareness. Their beautifully streamlined bodies could conceal frustrated minds. They may know that their water-based environment restrains exploration of the land, and having no hands with which to build could compound this irritation, if only because it seriously hampers constructive achievement.

A proper understanding of the manipulations that gave rise to our consciousness could have unbelievably far-reaching consequences, given the huge expanse of time remaining for life on Earth to continue. What now causes mankind, and possibly other species, to know that they exist is an extraordinary phenomenon.

It is extremely difficult physically to identify with any distinctive aspect of organic life in order to gain an understanding of mankind's awareness. It appears to be as hopeless as delving inside a modern computer to establish what creates its logic. So where should the search for the software and hardware of thinking processes begin, beyond that of the knowingness of species genes? It is a question that needs to be resolved. It cannot continue to be fudged or concealed within prettily designed phrases or poetic expressions.

8

The Cosmos

THIS SURVEY OF the field of scientific knowledge has now been completed. It was an essential first step in clarifying thoughts and objectives. To enable intuitions to mature, the memories gathered will now be recalled and reviewed. It is then hoped that careful examination and evaluation of these insights will lead eventually to the discovery of essential definitive truths. However, during this journey the United States National Science Foundation announced that new advances in cosmology were solving the space jigsaw puzzle. Before endeavouring to progress further, their press releases need to be scrutinised.

It was extensively reported that the riddle of the geometry of the universe had been solved, allowing exploration of its structure to be advanced. *Nature*, the prestigious science journal which publishes papers by eminent researchers from around the world, reported on data collected by the giant, balloon-borne Boomerang telescope floated twenty miles above Antarctica. The world was told that the universe in which mankind lives would expand forever. The telescope had obtained high-resolution images of cosmic background radiation, a telltale sign providing information about the very earliest history of the universe.

Scientists reported that the findings showed the geometry of the universe to be 'flat'. 'Flat' did not simply mean like a sheet of paper, it was rather that space is flat as opposed to curved. A flat universe demands that two lines starting off parallel will remain the same distance apart forever. The implication was that there is precisely the right amount of matter and energy to achieve a balance between the two possible curved geometries known as 'open' and 'closed'. A closed universe would have just about enough matter to reverse the big bang expansion and pull everything back at some time in the future into a big crunch. The parallel lines of the flat geometry mathematically portray a universe that will expand forever as cold, lifeless matter spreads out never-endingly at temperatures close to absolute zero. The role of mankind is therefore insignificant.

In contrast to the enthusiasm of the scientists, the popular press adopted a surprisingly cool stance. It was not just that the highly technical mathematical equations were open to interpretation. The suggestions that humanity counted for nothing, and that eternity would contain no people or angels, simply had little appeal. The response of the press expressed frustration: sarcasm became the order of the day. They showed little kindness to cosmology or its theoretical physicists. They protested that there were far more theories than facts, so it was possible for scientists to derive whatever theory they wanted from their mathematical expressions.

Readers of the daily papers were reminded that even Einstein had 'fiddled his equations' when he first presented his general theory of relativity to the world in 1915. By adding an arbitrary 'cosmological constant' to his calculations he had made them show a steady state – an unchanging universe – which was more to his liking. In fact the unadulterated equations had implied an expanding universe, as was later confirmed by astronomical observation. A dangerous precedent had been set. The press argued that this deception still weighs heavily on the consciences of all scientists following in Einstein's footsteps.

The announcement that a mathematical equation might now be agreed, explaining everything within the universe algebraically to the satisfaction of theoretical physicists, was met with scepticism by the press. The general public is not interested in an answer to the riddle of the universe that does not relate to their inner experiences. It is not an answer that mankind in general wants. It is not a proper answer. It is putting the cart before the horse. That there has been no recognition of this in the main thrust of cosmological thinking for three hundred years is difficult to understand. A change of direction and language is obviously needed.

That mankind is not happy about cosmology's progression is regrettable. However, the Science Foundation's report fulfilled a need for my quest. I was invigorated: it helped dismiss the 'what comes next' feeling that journeys such as this can so easily develop. Being reminded that all but five per cent of matter and energy in the universe is not understood or measurable was of particular help for it questioned Michelson's split light measurements, which were described at the time they were made as possibly the most important in the history of science. There is obviously a great deal 'out there' yet to be learnt.

It has not been easy to accept that the galaxies of the universe are moving away from the sun's solar system at about 125 miles every second. To realise that twelve billion years or more have passed since the big bang set the universe in motion is awesome, as is the fact that the original rocks of planet Earth are more than three billion years old. Within this recollection of the immensity of the universe there is now one intuition, one that appears and expands.

It is a visualisation of black holes that evolve in sizes ranging from the trivial to the enormous. One in particular, that possibly exists at the centre of the Milky Way galaxy, the home of planet Earth, has a mass considered to be equal to a hundred million stars – about one thousandth of all those in the galaxy. The image is of black holes which are recycling stations of universes which have become unsuitable for their purpose. Their 'flowers and seed pods' then need to be cut back, compressed and refined down again into their basic cosmic stuff. The whole process then commences again anywhere and everywhere within new and repaired universes. All are constrained gravitationally and emerge spontaneously to laws that maintain a balanced cosmos. With this image there is no longer the unacceptable notion of universes developing from nothing. But, if that is not the case, from what do they develop? What follows is a consideration of the utmost importance.

One is led back beyond this greatest of movements, from the activity of the universe to the cosmos itself. This can only be visualised as an eternal omnipresence. Such a concept may be difficult to accept, but the fact that alternatives such as 'nothingness' are far more obscure helps justify the realisation. At this stage it need not be accepted out of hand, but neither should it be rejected. It should be given due consideration; there is a need to see where it leads. Of course, whole universes are less difficult to visualise than the cosmos within which they unfold. This is because they have defined characteristics. They are created from big bangs and mature into atoms, stars and galaxies. They evolve cyclically within measurable time and have an end.

When considering the concept of the cosmos as a whole one needs to proceed with care. It is so easy to find that there is nowhere to go; but to argue the alternative, that creation just happened and that everything came into being at a point in time, is completely unsatisfactory. The obvious response had to be to question what was 'out there' before. The only logical way forward is to acknowledge that there is no alternative to the idea of the existence of these universes evolving within an eternal omnipresent cosmos, an ultimate reality. The understanding has to be that the cosmos itself cannot have had a beginning and will not cease to exist. An elusive 'nothingness' is all that can replace its eternal presence, and that has to be entirely unacceptable.

Thus, what can be regarded as a 'model of everything' had been found and it will be interesting to see how this eternal omnipresence stands up to scrutiny, exploration and development. Seemingly, the only possible structure has to be a cosmos that is indeterminate, within which incalculable numbers of universes cyclically spawn and die. But the ramifications need to be explored.

An important 'milestone intuition' relating to this exploration was the suggestion that the basic substructure of a universe emerging from its most primeval big bang state must be 'up quarks', 'down quarks' and 'electrons'. The reason for this deduction was a simple one. No evidence has been found to suggest that these particles derive from anything smaller. They are the essential 'bits' of atoms which are the only physical contribution that the cosmos makes to universes. But these big bang events cannot just happen. There has to be something out there that orchestrates them and from which they are created.

Atoms, the substructure of all universes, collectively and individually respond to the demands of cosmic management. These building blocks of absolutely everything within universes, these busy and tiniest of structures, are beginning to be seen as vital nerve responses. They interrelate like the thought patterns of consciously aware mankind and allied travel notes record another 'milestone intuition'. What else can such a concept be called? Tethered to gravitation's demands, these inorganic atom receptors ceaselessly respond in the same way as appendages of unaware organic life do when engaged in the business of survival and reproduction. Just as unaware organic life responds to its 'genome knowingness', so do the atoms of universes respond to the design inclinations of 'cosmic knowingness'.

There is a need for this realisation to become completely focused, and one line of enquiry is into the nature of organic molecules. Mankind does, of course, consider them to be special, but it may be important to decide exactly how they differ from inorganic structures.

For example, the remarkable way proteins self-assemble into machinery that is able to carry out very specific tasks with an optimum efficiency, indicates the operation of a remarkable intelligence. It is as if they appear to understand their own activity. They conform to the discipline of their function in ways that are otherwise difficult to understand. The behaviour of electrons in atoms and uncomplicated molecules has a similar quality. To put it simply, both appear to have identity and purpose. A milestone decision was the acceptance that a force of intelligence extends everywhere, intercommunicating with and regulating atom receptors in conformity with the demands of the cosmos. It is difficult to describe them other than as living, although inorganically. It has been suggested that all particles are part of a unitary quantum system, and it is now seen to be possible to envisage a single global wave function for an entire universe. Intuitively, it can be seen that the realities of all particles are interwoven with all others.

This introduction of an element of intelligence into the evolution of universes is not as radical as might be assumed. Organic life matured for billions of years in a similar manner and its knowingness has not been unduly questioned.

One of the difficulties in accepting an inherently intelligent cosmos is that, for a long time in our society, intelligence has been associated only with living things, and with mankind in particular. Only the behaviour patterns of living organisms appear to be individualistic and purposeful. In consequence, mankind's exploration of the source of intelligence has been directed only towards organic life. Inanimate matter has largely been dismissed as non-living and therefore devoid of inherent intelligence. It is merely used. It is sat on; bridges are built from it, as are other structures – not to mention complex devices for transport.

So why is it that organic life is looked on as special? At first sight it seems obvious that a living thing is superior to a mountain, a winding river or an ocean. It can run, walk and perhaps even smile and cry – but maybe such considerations should not be overplayed. The design of organic life's molecules has been discussed in detail. They consist of selective numbers of the same ninety naturally existing atoms that compose everything. Crude oil and coal were formed from decomposed organic matter, so they consist of the same atoms and continue to do so when they are burnt and become part of the inorganic realm. Nothing is actually lost and the atoms of the changed molecules continue to function ceaselessly.

Respiration is an absolute requirement for the mobile needs of animals but, as a chemical process, it is not peculiar to living things. In principle, if not in detail, it is the same process as the combustion of petrol in cars, the robots of humankind. The energy in car fuel comes originally from solar energy first held in plant life, then stored and concentrated through the slow anaerobic decomposition of vegetation. Nature has not designed animals that can obtain the sun's energy directly, or even from roadside pumps. Instead, their systems allow them to eat animals and plants. The required energy from the sun is then provided through the complex processes of digestion.

With these considerations the distinction between living and non-living matter starts to become blurred. The physicist Niels Bohr explored the distinction through complementariness: producing effects in concert different from those produced separately. A creature such as a rat can be observed as a living creature, and its behaviour as a whole organism can be studied. Equally it can be regarded as a chemical system, and the chemical processes within it may be studied. However, for this to happen the creature has to be cut up into small bits – and then it ceases to be a living creature. Bohr regarded the two ways of looking as complementary; you can use either, but not both at the same time. Therefore the conclusion to be drawn is that the distinction between living and non-living things is in the way they are looked at. It is in the eye of the beholder; it is not inherent in the things themselves.

The main component of all living things on Earth is water, so why should this substance not be regarded as an aspect of life? Similarly, it is now realised that the whole of the atmosphere is a product of living creatures and that all living creatures are utterly dependent upon it, so why should the molecules of air be regarded as less alive than those that compose a body? All are atomic.

The outcome of this line of thinking is that the attributes of awareness, the conscious appreciation of self, events and circumstances, has to be extended beyond aware organisms. It is impossible to differentiate between biological life's atoms and those of inanimate structures. The distinction is arbitrary. One of the peculiar findings of atomic physics is that, in certain subtle experiments, measurements carried out on one particle were found to affect those performed on other distant ones. Einstein was very sceptical about this 'spooky action at a distance', but, if each atom is a manifestation of an all-pervading intelligence, it might not be so odd. It is a reaction that results essentially from consciousness, albeit in a particular form or grade.

During this daunting cosmic journey I was frustrated by an inability to see anything that remotely related to a constructive explanation of conscious awareness, thought and sight. However, despite the fact that some intuitions depressingly implied that these phenomena were beyond human understanding, stubborn feelings persisted that all three occurrences were closely related and that an appreciation of any one of them would quickly lead to an understanding of them all. Indeed, an indeterminate picture of the cosmic whole that could provide a key to this understanding is now slowly forming.

These maturing thoughts now make it possible to accept that organic matter is just one aspect of a unitary whole. Organic life is distinctive though not cosmically crucial. If we can accept that other species may be genetically contrived to see and react to inner thoughts, perhaps it is possible to get used to the fact that consciousness is not peculiar to mankind but is an attribute shared with the cosmos itself. It is a thought that should be allowed to develop, as should the view that inorganic and organic matter are both inspired by the activity of atoms which are created from the intellectualised grain of the cosmos.

This point in the journey is well remembered. It had been accepted that the cosmos is eternal. Of equal importance was the conviction that, together with non-eternal universes, it 'lives'. Surprisingly, I began to relax, though it did seem necessary to take stock and decide whether these conclusions could be developed progressively or whether they were only of academic interest. Quite obviously fundamental issues are involved. When no consideration need be given to a cosmic beginning, the activity of thinking totally changes. The cosmos lives and has always done so. It is not necessary to be concerned with how organic life came into being, only to understand it to a greater degree and recognise it as a vital part of living universes.

It is inevitable that the complexities of organic life and the atom will, in time, be understood to a far greater degree than at present. However, it will have to be accepted that *everything* lives, although mankind's understanding of living will need to be redefined. A piece of granite has to be seen as a teeming mass of atom activity, as do solar systems, galaxies and universes. These working bits of the cosmic whole discipline multiplex worlds through cycles of chaos and order eternally. Universes evolve from big bang beginnings and end as a result of gravitational compression within black holes. When suitable environments are coaxed into existence in solar systems during their lifecycles, organic life is incubated as a result of cosmic law's opportunistic relationship with its atom sensors. The adaptability of the many species evolving towards consciously aware adulthood is then tested. It is a cosmic phenomenon that begins to generate 'feeling'.

Twenty-first century man has created and will continue to create many of planet Earth's problems. This is an important point to make at this early stage after the awesome experience of the journey. Mankind is complex in so many ways and for so many possible reasons. The diverse pictures conjured up during the journey may or may not be helpful when reflecting on ultimate cosmic goals. Whether they are issues that need to be considered further as the cosmic puzzle unfolds is questionable. It is just possible that the thoughts or pictures caused by talking about these matters should be 'retained within' and reflected on as the journey continues.

Thought now needs to focus on an inherently intelligent cosmos. It may envisage the creation of many advanced forms of sea- and land-based organic life to improve the chances of one consciously aware species surviving a cataclysmic event on land or at sea during the evolution of a planet. The meteorite collision that coincided with the extinction of dinosaurs and many other animals on planet Earth about sixty million years ago, leading to the age of mammals, is another possible exercise of cosmic intelligence. The recollection of this crucial event evoked the thought that the appearance of consciously aware land – and ocean – based animal species could be one of many eventualities that an eternally aware ubiquitous cosmos would accommodate within its universes. Their direction would be determined by inbuilt sensors related to its very psyche.

The big bang origin of the universe has now been widely accepted but it remains unconvincing as an explanation of its absolute origin. Even though sophisticated explanations involving quantum fluctuations have their appeal, it is questionable whether they really can overcome the simple philosophical premise that 'something cannot come from nothing.' The big bang theory deals only with those aspects of existence that are subject to change in time. It conforms to simple logic that there is an eternal presence, the cosmos, which remains unchanging whilst universes emerge from it and then return to it via the galactic 'laundering' that seems to be the function of black holes.

Its empowering aspect is cosmic grain which intelligently regulates the development of all inorganic and organic structures through cyclical processes. This is the very quintessence of matter, the thought patterns from which the physical atoms take form.

- **Being consciously aware invites questions.**
- **When an intuitive answer is not forthcoming, could it be because, at the cosmic level, the question is meaningless?**
- **'No' has to be the answer.**

9

Mitochondria and Chloroplasts

FROM A PHILOSOPHICAL point of view it is interesting to know that organic life is composed of infinitely more individual life forms than its cells. Though possibly aware of this, very few of us really think seriously about being multi-cell structures. The fact that each of them is extremely complex and lives a life of its own is given little consideration. There is then the fact that all are built from atoms, which this account has accepted are intellectually related to all others within the universe. Though not as yet mentioned, the microscopic-bodied mitochondria and chloroplasts now have to be accommodated: between twenty and many hundreds of one or other of them exist in every cell of a multi-cell organism, whether animal or plant.

These thoughts are very well remembered. At various stages of the journey frustration was experienced because, like many others, I have always distrusted these diminutive organisms. I suppose the reason for this is that I have always known that viruses can kill and are responsible for many well-known diseases. Indeed, one caused the 1918 influenza pandemic that killed an estimated fifteen million people. It was the largest number in the history of mankind to die as a result of this type of infection; it was double the number of those killed in the

Great War that had just ended. At that time no one knew what these tiny things were, the word *virus* being the Latin for 'poison'. A virus was viewed for the first time in 1944 under an electron microscope using an ingenious shadow casting arrangement. The cowpox virus was then seen to be barrel shaped and about 0.25 micrometers thick. One of the smallest ones so discovered was yellow fever, which is only 0.020 micrometers.

Too small to have very much metabolic machinery of their own, these feared pathogens use other living cells to reproduce. One within a human cell can become two hundred in less than half an hour. As parasites incapable of multiplying anywhere other than within living cells, they quite obviously could not be the original organic life form on Earth. Earlier single-cell bacteria do not share their dependence on living cells, and because they are complete organisms they are about a thousand times larger.

Apparently, medical doctors were resentful when first advised to wash their hands to protect against infections from these tiny complete organisms. However, they later learnt to respect the decision for the death rate in maternity wards alone fell from twelve per cent to less than two.

It is not surprising that thoughts related to micro-organisms are generally ones of revulsion; it is difficult to believe other than that they can kill and harm. Drugs are indeed used to destroy them before they kill us. It is therefore not easy to accept that mankind's every living cell is totally dependent on domesticated mitochondria. Indeed, all the cells of animals and plants are host to large numbers of these tiniest of bodies that have, in some way, been acquired. That is, unless they acquired the cells of all multi-cell conglomerates.

Disarrayed thought patterns have now converged. There are no more discords. To understand why, a more impersonal world must be imagined, one that was much earlier in the history of planet Earth. It was at a time when no plants or animals existed; only the most elemental organic life had emerged. The history of life patterns at this time is confused, for it was the period when the build-up of the planet's ozone layer gradually cut off ultra-violet radiation from the sun. Before it formed, simple life undoubtedly existed by breaking down complex chemicals into simpler ones and storing the energy released. This process required ultra-violet radiation to rebuild the needed chemical substances. Mitochondria-type single cells that contained chlorophyll to carry out photosynthesis must have evolved at this time. More advanced forms have since been named chloroplasts, and they utilise the sun's energy to exist. All are metabolic marvels that oxidise glucose with a net gain of up to thirty-six adenosine triphosphate molecules to one of glucose – cells without their sophisticated equipment gain as little as two.

When organisms became complex enough to leave fossil records these showed that animal cells lived by ingesting plant cells. Life was then a balancing act for survival, the odds very much favouring plants because of the abundance of energy from the sun. At some point in their evolution, single-cell or multi-cell plant life took on board chloroplast-type cells to power their metabolic needs through photosynthesis. It is possible that this was when their own ability to synthesise the sun's energy was being stretched to the limit by the collective demands of their ever-expanding multi-cell structures. Alternatively, it could have been a single-cell need.

I was initially frustrated by what seemed to be the improbability of this liaison, which history often records as though chloroplasts were acquired symbiotically. But time moved on and animal-type organisms were able to exist more freely as plant life became plentiful. There was a greater availability of vegetation for food. Just as plant energy problems were solved by domesticated chloroplast-type cells living within every single plant cell in numbers of up to forty per cell, an identical energy problem was resolved in animal cells. Mitochondria provided the required power in the form of ATP through oxidative phosphorylation, and they now dominate most animal structures. The number occupying each individual cell varies immensely depending on its particular requirements, such as active nutrient transportation. In consequence, the intercellular movements of these microscopic bodies within their multi-cell organism are considerable.

It is now generally acknowledged that advanced multi-cell animal and plant organisms could never have continued to grow in size without the availability of the extra energy provided by mitochondria and chloroplasts. Classifying them with other such partnerships is difficult, though many plants have been found to enlist help from an army of predator bodyguards such as insects, mites and spiders. They hide in the plant or are summoned from elsewhere by chemical signals. Plants have even been known to grow tiny pockets on the outsides of leaves to encourage their 'bodyguards' to stay. But the multi-cell 'in house' relationships with chloroplasts and mitochondria are quite different. Before these diverse symbiotic bonds were established, it would have been necessary for the proposed unions to have been envisioned and accepted, no matter at what level of evolution. Vision, intention and purpose were essential requirements.

The energy limitations of multi-cell structures would have had to be appreciated, and the important fact that symbiotic associations could solve the problem. That plant life required chloroplasts while mobile animals needed mitochondria would also have required a systemic understanding, as well as an acknowledgement of the need. This must be so, for it is surely impossible to accept that these could have evolved through trial and error. That such sequential events could have occured by chance on two entirely different occasions for different needs, possibly separated in time by tens of thousands of years, is not a reasonable assumption. It becomes an impossible one when consideration is given to the fact that the needs of both multi-cell plant and animal organisms had to be foreseen. An appreciation of the level to which multi-cell plant life had progressively to evolve before there would be sufficient growth to support the eating needs of mature multi-cell animals was also required.

That nature has done everything that mankind can think about is possibly a very profound realisation. It is certainly one that I found important. Associated intuitions somehow bring into focus those things that thoughts cause to be known. A cosmic intellect exists; there are now no lingering doubts whatsoever. Earlier reasoning to the effect that the eternal cosmos and its non-eternal universes live has also been strengthened. Indeed, the visualisation that the conglomerate whole is a cyclically disciplined intellection has been consolidated.

The complex symbiotic relationship that exist between chloroplasts or mitochondria and all cells, whether animal or plant, must surely confirm that 'vision' was required to appreciate and solve this multi-cell need. Not the vision of sight or pictorial symbols, but that of intelligence and foresight, the facility to conceive or know something – and the cosmos is its only possible source. No other choice is possible simply because both universes and planets die: they cannot have any retained vision. Perceptive guidance such as that allied to these multi-cell relationships of chloroplasts with plants and mitochondria with animals can only have derived from cosmic intellection.

It does, of course, have to be accepted as possible that the in-house relationships between multi-cell organisms and their respective microscopic organisms are not symbiotic. Arguably, they could be mechanistic. These tiniest of microscopic bodies could have been physiological designs that actually evolved within plant and animal cells. The need to increase oxidative phosphorylation to the levels required to support the increased demands of multi-cell life may have been evolutionary developments that were solved either way. However, it has to be remembered that the integrity of these tiny living structures is cunningly protected at embryonic conception, which is absolutely essential for this closely confined, long-standing relationship. It is achieved by allowing only the DNA within the female egg to survive all sexual unions. This continuing meticulous attention to detail in the planning of the cosmos must itself dismiss any lingering doubts as to its intelligence. It is too abundant to allow any other conclusion.

So, could it still be reasonable to argue that the absolute acceptance of a cosmic intellect is hasty, even naïve? The answer to this question must surely be 'No.' It has to be remembered that, although chloroplasts and mitochondria were the final catalyst, there is really no other satisfactory way of rationalising the disciplined behaviour of atoms. These building blocks of everything, these 'twanging guitar strings' that give no hint of their existence, could not possibly have just arrived. Add to this the complexity of the genes of species. From where could their vast programmed logic have derived other than from another intelligent source? When giving thought to these issues the intuitivism of the early Greeks should be remembered. Their historically recorded belief that organic life is fabricated from 'useable tiny bits' could not just have emerged. It is far more likely that these visualisations were the cosmically stimulated intuitions of a special few of this early civilisation.

10

Cosmic Grain

It NOW SEEMS IMPORTANT to return to the broader picture of Isaac Newton's era. It will be recalled that his inexplicable 'gravity' was considered to be central to all disciplines of the cosmos. However, in 1915 this changed with the introduction of Einstein's general theory of relativity, which resulted in its dismissal. It is surprising that the scientific world continues to believe that space is empty of this force because of Albert Michelson's split light measurements. To put it mildly, the dismissal of gravity would seem to have been more than a little hasty.

Cause alone is surely sufficient to argue that there is something 'out there', within and beyond the expanding universe: something that relates to the cosmic whole. 'Nothingness' is not an acceptable alternative, neither is Einstein's 'warping fabric of empty space'. Newton's theory of gravitation stated that the moon moves around the Earth because an inexplicable force of attraction pulls them together, and his mathematical equations unquestionably support this view. According to Einstein, the moon moves around the earth because the space in which it moves is curved. Newton's theory leaves little to consider intuitively while Einstein's claim is intuitively incomprehensible. Isaac Newton considered it absurd to believe other than that an agent acting constantly according to certain laws, whether 'material or immaterial', must be the cause of gravity. So perhaps consideration should initially be given to the scientific world's present beliefs.

The official announcement made by the US National Science Foundation when this résumé's cosmic exploration was nearing completion should initially be brought to mind. It will be remembered that its Boomerang telescope report convincingly stated that only five per cent of the mass and energy in the universe is detectable. Ninety-five per cent was then acknowledged to be immeasurable, deeply mysterious, dark energy and matter.

Even more recently, in the year 2003, the Wilkinson Microwave Anisotropy Probe, carrying high-precision instruments, raised again many of the questions published in the Boomerang telescope report. How were the stars born? What will happen to them? What is the make-up of the universe? What is mankind? These questions were just a few of the many that were put to Dimitris Nanopoulos with regard to the probe's discoveries. This holder of the Michell/Heep Chair of High Energy Physics, also a distinguished physics professor at the Texas A & M University, replied to the effect that the universe is a mixture of nothing. Talking about the universe, he more or less confirmed that it was expected to exist into eternity, continuously expanding. Discussing the experimental data gathered, he stated that this latest probe had shown images of occurrences 380,000 years after its big bang beginning which confirmed that, when the first atoms formed, energy changed from an undifferentiated soup of particles. He went on to say that the 'inflationary theory', first formulated twenty years ago, was upheld. The symmetry of forces was apparently then broken and anomalies appeared, one of them being us.

Dimitris Nanopoulos went on to say that fluctuations are considered to be essential if homogenous soup is to be avoided, so somewhere, quite by chance it seems, differences in the density of matter created the universe and, just as it appeared by accident, so did the first stars and galaxies. He added that snapshots taken deep within the space of the universe, when its age was a mere 380,000 years, apparently validated many things, including the fact that its expansion will continue forever because it is open and flat and does not have two dimensions. We are told that it follows Euclid's laws of geometry and is like an expanding cube. There isn't a curve anywhere within it, and nature chose this simplest course because the total sum of energy is nil. In consequence, the cosmological world now believes that the universe began from zero energy and mankind is a mixture of nothing, as is everything else.

When asked, 'What is matter?' the reply of Dimitris Nanopoulos was particularly interesting. It would seem that atoms and molecules, of which it is agreed everything in the universe is made, have once again been declared to constitute only four to five per cent of the universe. Twenty-three per cent is dark matter. Apparently it is dark because it emits no light and the prevailing theory is that it is possibly allied to particles with a net electric charge of zero. The remainder, which is more than seventy per cent, is dark energy, for which there is no prevailing theory or explanation.

It has to be admitted that this self-declared Wilkinson Microwave Anisotropy Probe certainty of hard facts is depressing. That it includes only the universe in which mankind lives is somewhat surprising. Confirmation that atoms account for only a very small part of the universe once again suggests that Michelson's and Einstein's 'space' is far from empty. Yet it will be remembered that 'empty space' resulted from Michelson's split light measurements, which made possible the dismissal of Newton's gravity and the introduction of Einstein's general theory of relativity. That so much of this space is now considered to be dark matter and energy importantly confirms the earlier Boomerang telescope findings. It also leaves no technical reason whatsoever to question an undetectable grain being introduced to the cosmos by this exploration.

Dark energy, the so-called 'cosmological constant', has now been heralded as one of the most important discoveries in science. This is difficult to understand because leading cosmologists must surely remember that long ago this 'ether' was acknowledged to be 'an all-pervading, infinitely elastic, mass-less medium' within which planet earth was immersed. As mentioned on numerous occasions, it was dismissed after Albert Michelson's split light measurements appeared to prove there was nothing outside planet Earth's atmosphere, thus providing a platform for Einstein's general theory of relativity. Presumably these things have been forgotten, as has Einstein's own conjecture that 'space bubbles with an invisible something'.

It is not too difficult to understand that earlier in this planet's history Isaac Newton questioned what gravity's all-pervading force could be. But even so, it was surely somewhat hasty for it all to be negated in such a cavalier way with the introduction of Einstein's relativity. There appears to be good reason to question, as daily newspapers have done, the irresponsibility of passing to the general public questionable cosmological data.

The object of this cosmic exploration is to establish truths relating to the extraordinary omnipresence that consciousness enables mankind to appreciate. The discoveries are not chosen ones or ones that were preconceived, but the route followed is worthwhile only if evidence cannot be found unequivocally to disprove or damage the intuitions enjoyed. Difficult questions have arisen that could possibly have cast doubts on the discoveries of this cosmic journey. The dismissal of Newton's gravity was a case in point, as was the introduction of Einstein's relativity. Criticisms of these were necessary because a watershed has to be negotiated. It therefore seemed appropriate to refer to these issues to verify an objective of this exposé, which is to legitimise 'cosmic grain'.

Because the cosmos is omnipresent, its elasticity can never cause its totality, the sum of all that is real, absolute and unchangeable, to become clumpy. It remains balanced, uniform and compliant, and its substance has to be seen as the root of everything, with whatever structured or unstructured condition such a concept demands. It is 'thought' which universes expand within and through, and the chaotic debris of worn-out stars is recycled by black holes of ageing universes. The reduced product of these cleansing machines is then reabsorbed into the substance of the eternal omnipresence of the cosmos.

'Cosmic stuff' wilfully excites the creation of time-orientated expanding universes. It then intellectively creates atoms from its own all-pervading substance within budding universes to commence the evolution of inorganic, galactic probabilities and possibilities. The well understood organic genomes of species also result from this activity and orchestrate the disciplines of animals and plants that evolve within solar systems, as and when suitable environments develop from these grains of thought that are the cosmos. Unquestionably, this cosmic substance, this incomprehensible, gravitationally inspired fabric of everything, is 'cosmic reality'.

As discussed earlier, Aristotle originally decided that the world consisted of an invisible outer shell of fire, as well as earth, water and air. The inclusion of fire was possibly because on occasions lightning could be seen flashing in the distant heavens. This envisioned world was, of course, very small and contained, but intuitive consciousness could have caused this early Greek philosopher to introduce ether somewhat later as a fifth element. At this time in mankind's history intuitive thoughts might even have pictured this 'outer shell' possessing a form of enlightenment, though it is highly improbable that cosmic intellect, as here envisioned, was a consideration. The proposed grain of the cosmos does, of course, validate the evolution of genes within organic cells. It is a part of life that has been discussed in considerable detail during this journey and the introduction of cosmic knowingness gives sense to this 'DNA equation', which must otherwise have been looked on as a totally inexplicable concept. But it is no longer mysterious: as the selectively encapsulated instructive data that dictate the unconscious behaviour patterns of organic species, which derive from cosmic grain, it does indeed become understandable.

This essential seed of everything is unquestionably a cosmic phenomenon. This has to be so, if only because there can be nothing beyond its eternal omnipresence. No apologies are made for the use of poetic nuances when saying it is that to which universes are reduced and from which they emerge. Enlightenment, intellection if preferred, is fundamental to this homogeneous amalgam. Universes inevitably follow its disciplines within the boundaries of its creativity. Some of this mankind has seen, understood, misunderstood, just admired, or accepted. Consciousness is also part of this cosmic experience, stimulated through its sight, hearing and thought. All these attributes arise naturally within species as brains mature to levels of complexity that allow the fine tuning required for closer harmonisation with cosmic reality.

In support of such radical assumptions it is important to talk more on this subject, simply because there is more to say. As the issue is intuitively based, it is hoped that the continuation of a relaxed conversational tone does not suggest an absence of what is often accepted as basic knowledge. Mathematical equations can easily be mistakenly looked on as reliable while intuitions are sorrowfully downgraded to indefinable thoughts of dreamers. 'Enlightenment of the inner man' should be afforded more respect, for it can, with considerable justification, be recognised as the fountainhead from which all knowingness derives. It is certainly a point of view that will be encouraged as this cosmic investigation progresses.

Cosmic grain, the fabric of the cosmos, will continue to be under this investigatory microscope. It controls and co-ordinates universes and their structures that it spawns and intellectually guides. Nothing could exist without the 'glue' of its complex interrelating omnipresence. It is a concept that would demand reservations had not this discovery trail encouraged the belief that scale and substance are elusively subtle. The differences between solids, force fields and even patterns of thought are confusingly indefinable. Large and small are simply relative measurements. Only the cosmos is absolute and investigations of the atom continually show this building block of universes to be substance, elusive force fields and inexplicable strings. Einstein called it 'spooky'; but cosmic grain is more, and it is necessary to justify this proposition.

Repeated questioning of Albert Michelson's interferometer measurements must be forgiven. The scientific world became convinced that all-encompassing 'ether' did not exist after his measurements relating to planet Earth's rotation were made. In consequence, Newtonian thinking was dismissed and absolute rest and motion were questioned. The era of relativity had arrived. Verification of cosmic grain necessitates further discussion: the neutrino seems to be the right point to begin.

Billions of tiny particles are ejected into space by the sun continuously. Neutrinos are one of the many types that pass unnoticed through planet earth and all its living forms. The realisation that they travel uninterrupted in this way and could pass through trillions of miles of lead without their motion being disturbed must again seriously question the validity of Albert Michelson's interferometer findings. But the grain of the cosmos is much more than tiny neutrino particles. The thought patterns of this grain respond instantaneously and, in so doing, they control atom activity within universes. Their relationship has to be a systemic one, for it is impossible for the substance of atoms to derive from anything other than the grain of cosmic omnipresence. There is nothing else.

Cosmic thoughts do not travel when imposing disciplines on universes, galaxies and organic life. They do not behave electromagnetically. They are static for the simple reason that there is nowhere for omnipresence to go. The force is mental and, as such, it behaves telepathically, as do the minds of the consciously aware when engaging in thoughts beyond the knowingness of genetic inheritance.

The grain of the cosmos is impossible to envisage other than as 'thought', the complex pictures of which are felt within. Translating these thoughts into words can be difficult, for language is not sufficiently developed to describe these inner experiences. To communicate mankind is confined to its vocabulary, and the inclination is to limit thought to suit language. There is indeed no other option unless it were possible to communicate intuitively. This profundity is the 'human dilemma'. It penetrates way beyond the superficial to the animating and vital principle credited with the facility of action and emotion. Many have named it the soul and refer to it spiritually. It cannot be seen or felt by any known detective device, yet its patterns envisage and opportunely guide conscious behaviour.

At no time has it been implied that the cosmos continually evolves. It is there. Its universes behave cyclically, and to that extent there are beginnings and endings, but essentially the whole is an eternal presence. In consequence, cause-and-effect behaviour patterns evolve repeatedly within universes when aided by the thought patterns of cosmic grain, which 'steer the ship'. No new knowledge can emerge but endless time insists that all universal behaviour patterns are repeated eternally.

There should be no difficulty in accepting that knowingness is storable. Cosmic disciplines create it in the genomes of living cells. Their chromosomally manipulated genes are obviously specifically designed so that stored knowledge evolves and dies with its organism. However, it is important to recognise that genes are pure information – knowingness tailored exactly to the same codes in all animals. It attends to the unaware demands of all organic species and this knowingness can derive only from the omnipresence of cosmic grain.

So what is cosmic grain? As structures of substance these grains are almost impossibly difficult for mankind fully to understand: 'tangible stuff' is not an essential requirement of something 'factual'. It is possible to go beyond solids to liquids and gasses; indeed the acceptance of atoms that are physically untouchable should help understanding. But it is important to go further than the seven to 735 bits that are built into the largest atoms, of which there are over one hundred types. It is necessary to bridge the gap between accepted substances and real, factual, authentic or genuine thoughts. The intuitions of cosmic grain can then be acknowledged as substance without mass. This grain contrives, engineers or manufactures universes through 'contemplation'. The knowingness of its grain accomplishes the objectives dictated. There is a need to feel this presence, which is to some degree analogous to that built into the genes of all organic life.

The lack of respect directed towards organic genes is quite remarkable.

The concealed, encoded knowingness of their chromosomal tapes extends to the equivalent of eight hundred bibles, data that would take a century to absorb if read at the rate of one word a second for eight hours a day.

On measuring this fund of knowledge against the amount that is absorbed in an average lifetime, it is difficult to understand how we can fail so miserably to show greater respect for what mankind *is* compared to what it feels it has achieved.

11

The Brain's Purpose

THOUSANDS OF DIFFERENT neurotransmitters, proteins and chemicals are required by man's brain to enable it to pursue its function. It is not therefore surprising that this very active organ generates a great deal of heat; but what exactly does it do? A point has been reached in this enquiry when it is necessary to know. Its cells are not greatly different from those of the rest of the living form, and it is accepted that their every-day needs are governed by genes. It is now important to clarify the degree to which this vital organ is responsible for behaviour and governance.

This cosmic journey has shown that many 'inner feelings' are largely concerned with day-to-day bodily functions. Hunger, pain, sexual desire and the urge for survival predominate; and all are known to be genetically determined. The information is duplicated within an organism's cells during embryonic growth. For humans it is stored on twenty-three chromosomal tapes, all of which are ceaselessly accessed to provide the information and controls needed for the working of cells, organs, and the body as a whole. The data that is required to assemble new animals and maintain them as living organisms throughout their lives is included, and there is little doubt that, after a complete organism has formed, a great deal of this genetic information is processed through the brain. This organ also plays a part in regulating the behaviour patterns of creatures such as in the navigational expertise of birds and the spider's ability to spin webs. In the higher animals, more than three billion years of easily recallable, genetically encoded knowingness is contained within every cell. This is not so obvious for mankind because there is an inclination to believe that much of what is known has been acquired. However, a very ordinary bird of greyish plumage beautifully illustrates the extent of this inherited knowledge.

Because female cuckoos lay their eggs in the nests of other birds, the newly hatched offspring do not experience any normal parental or other living guidance. Initially each one has to envision its own migration to the correct part of Africa, where it instructs itself on how to eat and find the right food. It then returns to its birthplace the following spring, acquires a mate and locates the nest of a suitable host bird in which it deposits its fertilised eggs. All this knowingness is genetically transmitted; no other source of acquisition is available.

The South African naturalist Eugene Marais, in his book *The Soul of the White Ant,* recorded a similar example. The weaver-bird builds its nest at the end of a drooping branch of a tree, such as the weeping willow, that grows near water. It is the most intricate affair, carefully woven and attached to the branch via complex knots. Although the adult is a seed-eater, it feeds its young on worms. In his experiment Marais hatched the eggs of yellow weaverbirds under canaries and repeated this for four generations. The birds were not allowed to build their usual nests. They were kept in artificial conditions: they never saw a piece of grass that could be used for nesting, or a worm, being fed an artificial diet. The fifth generation were released into their natural habitat at the right season and they began plaiting and building nests in the conventional way – they even built more nests than were required. What is more, when their young hatched they were fed on earthworms. Such is the extent of gene knowledge.

Similar genetic controls are exercised over the behaviour patterns of all organic life though they are less dramatically noticeable. Quite obviously catastrophic disruptions are successfully accommodated: a fund of adjustable instructive information has to be available to deal with the unusual and unexpected. To assume it has all first to be fed to the brain may not be accurate, for it is known that sea slugs can use their limited gene knowingness without a brain. However, the fact that vast numbers of unaware animals have brains must mean that this organ is needed to cause some genome knowingness to be directed to specified areas of the multi-cell conglomerate whole. A great deal of mankind's gene data must obviously be similarly attended to. But this is the preprogrammed gene data of each and every cell; it is not 'invented knowingness'.

If music can be described as selected sounds pleasingly arranged, then surely spoken language could be regarded as different noises designed to communicate. Interestingly, it is possible to represent both these sounds graphically, using musical notation and writing respectively. The documented information can then be decoded into the notes of music or the original words of speech. What is quite remarkable is that, no matter how or when the decoding occurs, the powerful emotional qualities music carries are revealed together with the form of the notes. In the same way, when black marks such as these are read, it is not just the sound of the words that is reproduced but also the various nuances of meaning. In fact, in silent reading the sound may be bypassed altogether, leaving only the meaning.

Of course, these sophisticated procedures relate to mankind's consciously aware state, but presumably memory in all living creatures works on a similar basis. The organic genome would then be seen to correspond to this 'graphics' phenomenon. It is indeed a software programme of detailed information, the complexity of which is beyond human comprehension. It can be argued that this is all unaware organic species are: a miracle of stored information cleverly choreographed by twenty amino acids of protein from a coded language based on three symbols expressing four alternatives. And this software is all contained on the chromosomal tapes of every living cell.

It is clear that written languages, musical notations and computer software change, develop and evolve with time according to altering needs. They have not come about by chance. The same surely must be true of organic genomes. They are intelligent contrivances; their organic software is a marvel of cosmic knowingness. Its accumulation of retrievable information has expanded to embrace all the evolutionary changes and demands of each species. It can be recalled as and when needed to control all the essential functions of individual organisms. Two issues arise from this.

Firstly, there is no reason why the genome cannot accommodate disruptions and be equipped to cope with very different circumstances. Something like a tax return document is able to embrace alternatives by giving instructions such as, 'if not applicable follow the red arrows and respond to the most appropriate question'. Generally speaking, similar programming procedures are used with modern computers. Secondly, although it is repeatedly stated that no computer can ever be equal in intelligence to the designer of its software, this is not an issue with the design of genome software. It never was designed, so there is no reason why genome software cannot become 'all knowing'. Complexity is not a problem. Genomes might be pro-grammed so that they eventually extend organic knowingness to the full potential of the cosmos.

Of course, the knowingness of the cosmos, unlike that of genes, cannot be located. There is no substance with which to relate. Its knowledge has to be whatever is necessary to service its universes and everything they contain. Its thoughts, judge-ments, knowledge and consciousness are all there is; its omnipresence determines this.

I am now convinced that consciously aware creatures have the possibility of direct access to the cosmic mind. They do not have to be limited to gene knowledge, extensive though it may be. Endeavouring to discern at what level living creatures connect to the cosmic mind is a crucial aspect of this exploration. Views have consolidated during the journey and it is now considered that progressive steps towards the conscious awareness that results from this union begin with the reflex nervous systems of lower order organisms. The maturity of the central nervous system is a key feature of more highly developed creatures. At the other extreme it is clear that mankind, the most highly evolved life form on this planet, has not as yet achieved full conscious awareness. There is further to go. Before the final step towards total awareness can be achieved, a greater understand-ing of the language of the cosmos is needed, and I suspect that this has to be acquired through intuitive thought patterns. This is because the cosmos does not have hands; neither does it have a voice.

While travelling within the cosmos I developed an intuition which led to the suggestion that a closer connection may be possible between the cosmic mind and organic genomes. The realisation related to the key purpose of the brain and central nervous system. It suddenly became clear that the brain was only a means of transferring knowingness; it is not a source of knowingness itself. That genomes transfer data from their inherited genes is conclusively known. Brains only receive this information, or knowledge, and distribute it throughout the organism serviced. They can thus be looked upon as organic computers responding to their own genetic software. It seemed illogical to assume that brains suddenly change into thinking organs when reacting to information unrelated to their genes. In many instances it is not even possible to identify when knowledge and the thinking process derive from genes and when they come from an entirely different source. In both cases the function is to collate information and direct instruction rather than to be the source of knowing.

There is a complication here that tends to cloud the issue. Humans can easily take a very personal view of the activities of their brains and identify strongly with what knowledge appears to be there. However, to assume that knowledge can be manufactured has to be presumptuous. It is patterned and applied as a result of usage but its origin, its coming into being, and the place from which it derives, has to be a cosmic phenomenon, whether it is designed into the genes of species or received in some other way. Its application is determined by the ability of organic brains to recall information that is available, the content of which has then to be usefully shaped into practicable and useable patterns.

Adapting to this understanding of knowing does, of course, cause problems. Is it reasonable to take individual credit for moments of inspiration when they are actually gifts from the mind of the cosmos? After all, who is this individual? It is a noticeable fact that all organic life expresses qualities of individual personality such as bravery, fearfulness or pleasure-seeking. These qualities eventually extend to the dynamic character, self or psyche that constitutes and animates individual human beings, making their experiences of life unique. However, it has to be obvious that individuality cannot be attributed to conscious awareness.

Uniqueness must surely be inspired genetically simply because unaware species are also distinctive.

12

Thoughts Become Intuitions

ADDITIONAL CONSIDERATION now needs to be given to sight, the intellect, the genomes of species and their brains. A particular wish of mine is to discuss whether a closer bond between man and the cosmos might be possible. It is an extremely interesting thought if mankind desires real fulfilment from its privileged position on planet Earth. There are in fact many points to be discussed, including life itself. As yet it has not been clearly defined.

In living cells knowledge is not received directly and immediately from the cosmos for it is held in genomes where it has slowly accumulated during the processes of evolution, like data in a computer. In man-made devices it is recognised that microprocessors are where the essential processing of data takes place. Other components exist only to bridge gaps between the operator and the processor. These complex switching devices suitably manipulate data requests from operators before passing them on. The brains of living creatures fulfil the same purpose: they are not thinking organs, a fact that bears repeating. These 'organic processors' respond in exactly the same way to the stored information of multi-cell genomes as do the more familiar silicon microchip processors to commercially available software. Both are sophisticated switching devices that shift, transfer, divert and exchange, or cause the operation of data to begin or cease. In the case of organic nervous systems, it is to cause conglomerate life forms to respond to external stimuli via their in-built instructions.

Complex silicon chip processors often accommodate millions of transistors. However, despite such technological achievements, present-day computer systems are crude when compared to what nature has produced. Organic gene software and its hardware processor, the brain, are built with atoms: the engineering is orders of magnitude smaller than even the finest silicon microchip that is theoretically possible. It is remarkable that fully operational genes exist in even the simplest life forms, with exactly the same system in all creatures, whether animal, plant or bacterium. Only the actual information stored varies within species. The programming also appears to be designed to develop with the demands of individual species. This underlying intelligence, that makes complex animals perform in the way they do internally and externally, is surely much more of a requirement in a living creature than in a man-made computer.

For consciously aware creatures such as mankind there is the possibility of a more immediate and direct access to the knowingness of the cosmos. The machinery for this seems to be the same, namely the combination of brain and nervous system with organic genome data. It is completely unnecessary to argue that purposes or functions change selectively when receiving information or thoughts from cosmic grain rather than from genes.

The consequence of this is that it becomes necessary to dismiss the brain as a thinking organ that invents its own enlightenment. To get a sense of the procedures involved when brains of advanced species absorb information from cosmic grain, 'the data bank of all knowingness', it may be helpful to consider the inner process of 'searching for thoughts'. To some degree this searching must be similar to the activity employed when individual organisms receive data from their genetic software. There is of course no 'within the body contact' but the many forged lines of communication along which genome language travels are available, and cosmic grain is 'the basic stuff of everything' including these genome highways. Perhaps the changes that take place in species that give rise to conscious awareness are such as make these same lines of communication available to receive cosmic grain's knowingness directly. It is a matter that quite obviously needs further consideration.

It is important again to acknowledge that brains do not think. They could be likened to radio receivers. Although the voices that speak with intelligence, meaning and wit appear to be coming from within the box, everyone knows that the real source is else-where and that the job of the radio is to tune in to the waves that emanate from that source and then translate them into a form that the listeners appreciate and understand. The design and connectivity of brains could be envisaged likewise as having the capacity to look outward or within for sense impressions, wherever they might be, when seeking applicable data that has relevance. They are 'search machines' that establish routes to desired data that is then suitably identified and selectively linked. This knowingness does not necessarily remain within an organism throughout its life, as is the case with gene data. These established lines of access to recognisable data should be asso-ciated with such adjectives as intelligent, gifted, artistic and pleasant.

Endeavouring to reach into the 'mind of the cosmos' to explore the way that it works and thinks was originally a 'fun' thought related to earlier cosmic considerations. However, I now admit that more recent intuitive thoughts have indicated that actually it would be difficult to imagine a more exciting adventure. Different strategies have presented themselves, which could be explored. One way is through attempts to contact other life forms elsewhere in the universe – and for some the wish to communicate with intelligent extra-terrestrials is quite an urgent one. In consequence a great deal of time and effort has been spent sending carefully designed radio signals, that could be seen to have originated from an intellectual source, in all possible directions.

There have been reports that planet Earth has been visited by extra-terrestrial beings. Many male and female humans claim that such aliens abducted them, after which they were examined and returned to their homeland unharmed but puzzled. These experiences will continue to be reported as long as the sense remains that we are not alone in the enormous expanse of the universe.

There is another possible route that could lead to a connection with cosmic knowingness, comfortably and relatively inexpensively. The genomes of advanced organic species might have been designed to reveal aspects of cosmic truth at certain levels of consciousness. This extraordinary 'book of instructions' that can copy and read itself, these twenty-three separate pairs of chromosomes of mankind that contain sixty to eighty thousand genes encapsulating an unimaginable amount of selective information, could have a purpose beyond just the regulation of cells. Conceivably, it could contain an inborn cosmic message. Genome language is known to consist of three-letter words formed from a total of only four letters, all of which is contained in long chains of molecules. A great deal has already been learnt from reading this form of knowingness. However, it seems possible that mechanistically focused studies within and beyond essential physical genetic information could reveal a concealed understanding of cosmic behaviour. It is possible that a great deal could be learnt from research in directions other than the known purpose of genes – concerning repressors and enzymes, for example. After all, the purposes of large portions of DNA, the so-called 'junk DNA', are still completely unknown.

Although only a small part of the omnipresent cosmos, mankind's intuitions can not do other than reflect cosmic knowingness. Its possible desire for intellectual closeness is therefore understandable and might also be a cosmic objective that embraces all aware species. There surely has to be an intuitively recognisable harmony simply because the 'bits' that identify mankind, including its genes, enjoy cosmic grain's systemic bond that relates everything to everything else. It caused mankind's genes to be the most sophisticated logic 'chips' on planet Earth.

It is surprising that direct communication with the cosmos has never been seriously envisioned. It is odd that it has not been looked on as an option. Researching the human genome beyond the needs of medicine could possibly be extremely rewarding, for direct communication in any other way would be difficult. The cosmos has no voice. Motivation can, of course, only follow the conviction that an enlightened cosmic message would be infinitely more likely to yield desirable results than the present-day interplanetary space adventures. The ultimate aim is the same: to gain access to total enlightenment. These were my 'flashpoint' intuitions that resulted in an initial feeling for the subject.

To participate in such a venture, consideration would need to be given to how cosmic planning would envision a consciously aware species approaching the issue. Such a species would obviously be nearing the final stages of evolution, and perhaps designed messages within its genomes would need to have been choreographed so that discovery could be achieved only by intellects that were sufficiently mature. They could be symbols or even releasable feelings cunningly concealed.

Could mankind persuade itself to engage in such an adventurous undertaking? Might the inspiration to search for a cosmic message be fostered? Would it be possible to convince world leaders to the degree that finance was made available? In other words, is mankind yet sufficiently mature intellectually, temperamentally and sociologically to respond?

To look for doors that lead to total knowingness, possibly to find and understand a message in an intuitively expressed language of mentally sculptured images with keys secreted within its genome, must surely stretch ingenuity as never before. Might mankind intelligently engage in what was originally a fun-thought adventure?

Conscious awareness and thought quite obviously remain frustratingly inexplicable and I am reluctant to discuss them as yet, simply because other issues need to be clarified before a sufficiently mature analysis is possible. One such issue is sight, for there is now no doubt that cosmic grain shares this extraordinary gift. Its enactment is, however, confined to universes. It is known that the eyes of organisms direct the mental pictures of universes into the inner molecules of seeing species. Without these precious attributes a tough hide and skull could possibly shield all against the force fields of universes that during these travels have been shown to cause 'seeing'. Their orientation allows genome calculations to instruct sighted species where stimulations are in relation to themselves.

Simple forms of animal life 'see'; the development of sight does not need an advanced intellect. It is not a big deal at the cosmic level. But the act remains something that needs to be understood. A continuing mystery is what exactly this image or feeling within is; but, whatever it might be, the activity of atoms predominates and those of animal organisms obey similar regulations as those in the inorganic world. There is no evidence to suggest that organic molecular structures are needed for 'seeing'. They have been shown to be an obvious necessity in the design of self-reproducing animals, but that is all.

It is true to say that the cosmos sees its domain within all universes through its own omnipresent grain. The same causes allow organic life to see directionally through orientating eyes. The process of seeing does not change; it is just that the eyes of animal life focus on data that is required to fulfil their mobile needs. The act relates to the cosmic grain of atom structures being excited by electromagnetic waves. Whether they are organic or inorganic is incidental. It is cosmic grain that sees and this intellectual force is everywhere. Absolutely everything derives from it and has the possibility of seeing what is within universes, whether it is free or imprisoned within structures.

Though not inventible, total enlightenment is eternally available everywhere. Regrettably it is not readily understandable.

All organic seeing is activated by reflected electromagnetic wave stimulation of its systematised cosmic grain. This directional seeing derives from complex gene calculations of eye orientations.

**How
        different
are we?**

13

Mankind's Neighbours

DOES MANKIND GIVE consideration to the degree
that it has imposed its self-importance on planet Earth? The
extent to which the planet continues to be exploited suggests
not. Consciousness, that allows the promotion of these thoughts,
does of course cause related anxieties. It is in fact a classical
chicken-and-egg conundrum of the type that the fifth century
Ionian philosopher Zeno specialised in. An improved understanding of the 'thoughts within' that make us what we are may,
in time, allow our species to decide if it can, cannot, will, or will
not evaluate such issues; whether it will live in harmony with
the cosmos or pursue its own ends.

Meanwhile, perhaps a survey of the progression of organic life
would be helpful. It might possibly establish whether it can be
sensibly suggested that the evolutionary routes of the cosmos are
designed towards particular ends. What then if cosmic aims are
found to be different from those that mankind is directing itself
towards? Such a conflict of interests could obviously create problems. Maximum fulfilment can surely only be achieved when the
consciously aware adults of a species direct themselves towards
the laws of cosmic behaviour. If this is not the case, consideration must surely be given by that species to changing the ways
in which it conducts itself.

Put differently, forceful arguments have been presented to the effect that universes, galaxies and solar systems are intelligently guided in preferred cosmic directions. That these designs evolve into entirely suitable planets within which organic life can progressively mature could be considered to be a cosmic objective. An advanced aim would be to guide genetically controlled organisms to evolve into species that are consciously aware. Is the evidence consistent with the proposition that such patterns are preferred or even mandatory cosmic objectives? Might these goals be built into its very structure? In this regard it will be helpful to survey the progression of organic life on planet Earth to see what this shows.

Fossilised remains preserved in some of the very oldest geological layers provide evidence of when the first living organisms appeared. These were bacteria and blue-green algae, now called cyanobacteria. Radioactive dating has established that their origin was more than three thousand million years ago. Even simpler forms of life must have originated much earlier but no trace of them has yet been found. Scientists have had to use extrapolations to postulate what LUCA, the 'last universal common ancestor' of all present life forms must have been like before diversification began.

The evolution of more complex organisms followed as an oxygen atmosphere formed, from the simple prokaryotes to eukaryotes, still single-celled but now with organelles such as chloroplasts and mitochondria. Multi-cell creatures followed later. The few that have left fossilised remains show very simple body patterns. Some bear resemblances to jellyfish.

The Palaeozoic Era started about 570 million years ago and covers events over the following 290 million years. The Cambrian was the earliest period of this era and it is probable that only water-based life existed at that time. This included marine coelenterates, widely diverse invertebrates characterised by radially symmetrical bodies with sac-like internal cavities. They are the corals, molluscs, anthropods and worms of planet Earth. The Burgess Shales, discovered in British Columbia at the end of the last century, give a very full record of the diverse fauna of that time. Because many of these creatures had shells, this period marks the beginning of a distinct fossil record.

A hundred and thirty million years into the Palaeozoic Era, during the Silurian Period, fish-like vertebrates first evolved. Here was to be found the early development of the complex internal vertebrate structure composed of bone and cartilage which protects and supports the soft organs, tissues and other parts of advanced animals.

Two hundred and eighty million years ago, during the Permian Period, there were significant developments and specialisations amongst reptiles, particularly mammal-like reptiles. Their full development extended over the following 150 million years, during which time they ruled over land, air and water. The period ended with the greatest mass extinction recorded in the fossil record. This marked the end of the Palaeozoic Era and the start of the Mesozoic.

During the Triassic Period which followed, dinosaurs, plesiosaurs and ichthyosaurs arrived on the evolutionary scene. They reached their maximum development in the Cretaceous Period, followed by their rather abrupt extinction about 65 million years ago. This marked the end of the Mesozoic Era.

The Tertiary Period which followed saw the ascendancy of birds and mammals. The mammals had slowly developed from reptiles back in the Permian times, and birds from dinosaurs in the Jurassic. It was only after the demise of the dinosaurs that they came to predominate.

The Quaternary Period began a few days ago in evolutionary

terms. Its two epochs are the present-day Holocene and the earlier Pleistocene, which extend over one and a half million years. These most recent times have been characterised by rapid climatic fluctuations, with periods of considerable glaciation interspersed with warmer spells. It was during this time, and perhaps a little earlier in the Pliocene, that the genus *Homo* appeared. The age of man had commenced.

The progression of organic life on this planet ought to convince mankind that caution needs to be exercised when assessing its own level of importance. It is only necessary to compare the diverse communities that have been 'at home' on Earth for hundreds of millions of years to the recent introduction of *Homo sapiens* within the last 150,000 years. These timings make it extremely difficult to establish any exclusive right to ownership. But more needs to be known before a judgement is made. The previous occupants of our planet have been briefly surveyed. Who are the present incumbents?

The simplest multi-cellular animals are the porifera or sponges and there are about five thousand existing species. They range in size from small to about two metres wide. Some live singly while others join together to form flat incrustations. All live in seas or fresh waters. Their jelly-like bodies contain a variety of cells necessary for organising life and their skeletal structure has many small openings through which water enters. The food of sponges is plankton, which is trapped as water passes through the body of the sponge. Reproduction is by means of eggs from which larvae hatch.

The Coelenterates include jellyfish, the Portuguese Man o' War, sea anemones and coral. They have the simplest of nervous systems and are not much more highly developed than colonies of single-celled creatures.

Worms have nervous, excretory and reproductive structures. Size varies from a few millimetres to about eighty and they have no respiratory or vascular systems. Even so, they can be free living or parasitic, and all produce vast quantities of eggs. There are more than 25,000 species.

Molluscs vary greatly in size from small shellfish to giant squid which are often twenty metres in length. Typically, their bodies are soft and slippery because small glands on the skin secrete mucus. They consist of a foot, the visceral mass, and the mantle. The foot is muscular and it supports movement that is aided by contracting and relaxing muscles. Most molluscs have a protective shell, their mantle, into which they can fully retract. Classical shapes are the numerous and well-known spirals and the paired bivalve or two-halved shells. All shells grow with a body that has well-developed organs: mouths that can tear up or grind food and a manipulative tongue are well formed; food passes to a stomach and is excreted through an anus, and there is a competent nervous system. Sexes are usually separate, and there are at least 130,000 recognised species. Only arthropods in the animal kingdom have more.

Segmented worms range from small to about three metres long. The first segment is the prostomium and the last the pygidum. In between there can be several hundred segments and the digestive and circulatory systems extend throughout the whole body. Blood can include red haemoglobin or the green compound chlorocruorin. There are around 9,000 species.

As already mentioned, the most extensive phylum is that of the arthropod. The number of species is in the hundreds of thousands. Its members usually range in size from tenths of a millimetre to several centimetres long, though the largest can measure up to three metres. They include scorpions, spiders, mites, crustaceans and insects. The body is covered with a cuticle containing chitin. It is segmented, as are its appendages, which allows considerable movement in the legs. The vast majority lay eggs, though a few species give birth to active young.

Echinodermata are remarkable for their evolutionary age. Their radical body symmetry is an important characteristic in so much as it has five arms that are positioned under a thin epidermis, with the needles of the skeleton projecting through or from this skin. Many are beautifully coloured. They include sea lilies, starfish, sea urchins and sea cucumbers.

The chordates are a very large and diverse group. What they have in common is a skeletal rod that forms a supporting axis for the body and protects the central nervous system. There are aquatic and terrestrial species of all sizes, ranging from very small to the largest animals ever to inhabit Earth. At present there are 45,000 species divided into three sub-phyla, several classes and a number of orders. The first two sub phylas include small marine animals while the third, and by far the largest, the vertebrates, contains the most highly developed and complex marine, freshwater and terrestrial animals.

Their bodies consist of a head, trunk and tail. The notochord extends into the middle of the body during the embryonic stages, after which it is replaced by a spinal column and backbone. There is a skeleton of bone and cartilage, usually with two pairs of limbs. The range extends from fish and amphibians to reptiles, as well as birds and mammals. The eggs of fishes and amphibians do not develop in water: instead they are enclosed in embryonic membranes. The main classes are as follows.

- Agnatha or jawless fish are aquatic vertebrates having snake- or eel-like bodies covered with smooth skin. A single continuous fin runs down the length of the body and there are only the rudiments of vertebra. Sexes are separate and larvae hatch from eggs. Only 30 to 40 species exist.

- Cartilaginous sharks and rays have bodies that are covered with hard, sharp placoid teeth-like scales. Five to seven pairs of gills behind the head allow water to flow through the gill chambers. The tail fin has the largest lobe on top. The heart is two-chambered, with a spiral valve located in the gut. There are more than 200 species.

- The great majority of existing fishes have skeletons that are predominantly bony, and they breathe through gills. Bodies are usually covered with scales, but occasionally the skin is bare or bony plated. The head ends in a snout and the eyes are on either side of the head, with one or two pairs of nostrils in front of the eyes. Behind the head is a single gill cleft. Pelvic and pectoral fins attend to locomotion. The fertilisation of eggs, produced in large numbers, nearly always takes place outside the body. There are 25,000 species.

- Amphibians usually have a bare skin with great numbers of glands. There are two pairs of limbs. They breathe by means of lungs, as well as through the skin. Fertilisation can take place outside or inside the body. The eggs produce tadpoles that develop legs as they grow, and after metamorphosis most amphibians live on land. There are approximately 2,000 species.

- Reptiles are vertebrates, some of which have reverted to water. Reproduction is by means of eggs which develop on land. The horny skin is in the form of scales and plates. In some groups the limbs have become atrophied. About 4,000-5,000 species exist.

- Somewhat surprisingly, turtles, tortoises and terrapins are reptiles with short bodies encased in hard bony shells that are flat underneath, with openings for the head, legs and tail.

The jaws are toothless. There are aquatic, terrestrial, carnivorous and herbivorous types, of which there are about 250 species.

- The bodies of birds have a high constant temperature and they are covered with feathers. The forelimbs have been modified into wings and the jaws are much elongated and covered with a horny sheath. The embryo is encased in an embryonic membrane that is usually referred to as an egg. There are approximately 10,000 species and they can be birds that run, birds that swim, or birds that fly.
- Mammals, like birds, have a constant body temperature. Their central nervous system is well developed, and the body is usually hairy. The two pairs of limbs are adapted for walking, climbing and burrowing. Sexes are separate and the reproductive organs are markedly different from all others, though the lowest group lays eggs. Placental mammals develop in the womb, where they are nourished. They feed on milk produced by the mother. The ears are sophisticated, and the head is joined to the spine by two occipital condyles. There are about 8,000 species, all of which suckle their young.

So, can any of this catalogued data of the Earth's fauna help resolve puzzles exposed during this cosmic journey? The immediate first response has to be that this short biological tour has graphically re-emphasised the fact that it is impossible to claim to be individual. All animals, including the species *Homo sapiens*, are part of one family. We all are descended from common ancestors.

There is then the fact that the earliest bacteria are still thriving, though they originated in the Archaeozoic Era, 3,500 million years back into our planet's most ancient history. It is frustrating that the significance of these simplest of life forms is so poorly understood by today's antiseptic world. As already stated, the tendency is to look on them as enemies that have to be destroyed, yet it is known that this is as ignorant as it is dangerous, for all multi-cell structures totally depend on a vast moving population of bacteria. The belief that intellectually mankind is superior to other species may be justifiable, but the mitochondria that make this existence possible are cells in their own right. Their thinking is obviously different from that of mankind, but without their genetic thoughts, that allow them to synthesise the complex and vital needs of species, all multi cell-animals would die immediately.

The question of species and their 'evolutionary bumpy ride' needs consideration. Why have so many hiccups occurred, a third of the way through Earth's useful life? It is arguable that species must take time to evolve on new planets, but why so many and why monstrous dinosaurs and whales? A great deal of thought has been given to this matter and an intuitively acceptable reason could be that designs are delivered in identical standard packages to all the emerging planets of universes.

A second more philosophical response relates to an impression that an unnecessary amount of time was squandered between the big bang beginning of the universe and the emergence of organic life on this planet. Because everything in the universe is anchored to the same expanding skin or fabric, the difference in time between planet Earth, and possibly many others, becoming suitable homes for organic life is probably insignificant. In terms of 'three score years and ten' it might be seen differently, but this cannot be a fair yardstick for measuring universal behaviour. So, does the progression of life towards 'significant species' in universes of the cosmos really require the long and bumpy ride suggested? Organic life's evolving structures are, of course, different from the progression of stars and their satellites, so to talk about a long time must be meaningless. Everything surely has to relate to everything else for it has been agreed that, within this eternal cosmos, everything lives, including galaxies, solar systems and their planets.

The sun of our planet's solar system began its life five billion years ago, and organic life emerged on Earth one and a half billion years later. Our planet will have died within the next ten billion years, as will its sun, so it is obvious that the progression of organic life on Earth can take a leisurely route to fulfilment. This will be so whatever organic species cosmic law decides will fulfil its planetary aims. Time measurements also suggest that it is reasonable to accept that the fabric of the universe has only recently reached the level of maturity necessary for existing planets of spawned solar systems to incubate appropriate organic life. In consequence, mature life on Earth could not have been expected before the Archaeozoic Era.

On reflection, it has to be logical to argue that Earth's timetable of events would roughly approximate to the timetable for organic species to be spawned anywhere within this universe. It also seems likely that there is a sufficiently long period ahead for such events to be completely re-enacted from start to finish before the whole structure dies.

Widely diversified planetary environments within the omnipresent cosmos must be very demanding but, if intellectual progression is a cosmic aim, it is extremely difficult to imagine dinosaurs serving a useful purpose on this planet. An interesting point is the puzzling fact that they became extinct suddenly, possibly from the disastrous climatic effects of a meteorite known to have crashed into the planet around that time. Yet they had successfully survived for more than 150 million years. Might the design of an all-seeing cosmos be such that it was acknowledged that dinosaurs had nowhere to go on this particular planet? Could the same decision have been made about whales? Like the dinosaur, they are massive. They are also a seriously endangered species because of over-fishing by mankind. It could be possible that inbuilt laws of cosmic behaviour wilfully protect the integrity of the cosmos when necessary. Mankind's determination to over-fish whales could have been excited cosmically after dolphins had been accepted as the better sea-based insurance policy against mankind's demise. For similar reasons the balance of cosmic intellectual controls that protect progressive species from disastrous meteorite collisions could have been eased while the problem of redundant dinosaurs existed. It would be interesting to know whether cosmic law is designed to protect mindfully in this way.

Dinosaurs became 'unfashionable' at the end of Mesozoic Era. Following their disappearance mammals and birds were able to radiate, and the flora and fauna that are familiar to all began to cover the planet. Mammals are the most sophisticated group of vertebrates known and, being warm-blooded, they are more able than others to endure changes in climate on this planet. They have acquired the ability to remain at 36° to 39°C whether in a hot desert or an extreme polar region. All but the very primitive mammal species protect their unborn embryos within the mother's body, where nourishment is received through the placenta, and when born they are suckled on the mother's milk. There are three to ten million species, the wide range partly reflecting differences of opinion on whether a particular animal is considered to be a species or a sub-species.

Species vary from those weighing very little, and measuring a few centimetres, to the most massive great whales weighing around 150 tons. Though extremely diverse in appearance, they are all characterised by having highly developed brains.

It is widely believed that the mammal species *Homo sapiens* first appeared on Earth during the late Pleistocene. Discoveries of traces of mitochondria in the cells of its females have made research into their earliest movements possible. As mentioned earlier, the bacterial fuel source of male cells does not pass to the female cell at conception, thereby giving protection to the integrity of the female DNA. Because of this it has been possible painstakingly to follow the movements of these earliest of humans.

It appears that they were determined hunters. Discoveries of the remains of their distinctive stone tools and weapons have provided further clues to enable a picture of their way of life to be built up. Evidence suggests that they were able to recognise the seasons and also that there was an oral tradition of learning, with stories passed down through the generations. Fossils and skeletal parts of ancestral species such as *Homo erectus* that predate *Homo sapiens* have also been discovered. It appears that this and other early species failed to adapt and eventually disappeared.

A currently accepted theory is that the original location of *Homo sapiens* was Africa. A hundred thousand or more left this homeland about eighty thousand years ago by first crossing to Yemen. From finds of their artwork a path has been traced back more than thirty thousand years, showing how they eventually spread to populate the entire Earth.

What intuitions have emerged from these observations about species, purposes and objectives? It is suggested that species follow similar blueprinted evolutionary trails on all the inhabitable planets that can sustain life. Too many species on Earth have failed or had little purpose for any other arrangement to be considered. All universes do of course start with identical big bangs, and then spawn similar bits and pieces to create atoms and molecules. But observation strongly suggests that the requirements of organic life cannot be identical for all planets. Differences in galaxies and their structures of stars and solar systems are many and varied, giving rise to different conditions where life could arise. The bewildering numbers of life forms that have evolved on this planet can surely only be needed if cosmic planning has one master plan for all eventualities.

14

Mankind Scrutinised

WITHOUT QUESTION man is only too ready to produce formidable arguments to justify his self-imposed image of supremacy on planet Earth. However, this journey has exposed doubts as to their validity. His dexterity and intelligence certainly ensured that he would copy the planet's secrets and use all available natural resources to best advantage. Regrettably, in the quest to do so, he has exposed the planet to possibly cataclysmic global warming and pollution, including serious danger from radioactive waste. In consequence, Earth itself might not survive as a healthy land-based home for organic life over a sufficiently long period for its present evolutionary cycle to be completed.

It has been argued that man's superior ability to reason results from the advantages of walking erect and having extraordinary gifts of manipulation. Using these criteria it would be difficult meaningfully to compare humans with dolphins. They have neither of these attributes, yet many believe them to be consciously aware, and it is accepted by some that their intellectual potential could be greater than that of mankind. The many creature types that could exist in the wide-ranging conditions of emerging planets may also be determined by a cosmic need to duplicate advanced species evolving on specific planets to procure the advantages of both land and sea-based environments. Having both dolphins and mankind would provide for the possibility of protecting against a cataclysmic failure within one or other of these two environments. However, it also has to be acknowledged that dolphins would be available should mankind fail to achieve cosmic adulthood.

Considering the evidence, my intuition is that no species is necessarily special in a truly cosmic sense. A vast range can be subjected to intellectual guidance if necessary, and one could easily believe that cosmic concern might well be justifiable in relation to mankind's exploitation of the planet. When encapsulated within its present level of sophistication, enjoying the cosmos that its conscious awareness and knowingness allows, the need to impress competitively surely has to be seen as somewhat absurd. It is possible that dolphins might have a closer bond with cosmic reality at this point in their evolution. It is an issue that should be soberly evaluated by our species.

Irrational competitiveness is worrying. It is clear that the demands of mankind on its environment surely go way beyond the level needed to ensure its progressive integrity. World leaders are invariably severely contaminated with ideas based on competition and their personalities can so easily be distorted to those of opponents, rivals or adversaries. Instant economic gratification always seems to be their popular selling point. The philosophy of the administrations, supposedly serving the different nation states and other economic entities, too often becomes 'spend ourselves rich'. They invariably persuade themselves that their particular vote-winning brand will put an end to recessions and all the dangerous sociological consequences of mass unemployment. The population's revulsion at inflation is also used to good effect by those seeking to retain leadership.

They repeatedly convince themselves, and the mass of people, that this problem can be overcome in some easy and popular way. It is unfortunate that national debts continue to soar unchecked when new money printed to finance massive deficits goes largely unnoticed. Unquestionably, if governments were corporations most would be bankrupt. It is impossible for such vote-catching operations to continue indefinitely. Inflation is, of course, the price that is paid for such deception, with all the ugly repercussions that come in its wake. This begins with recession. Unemployment then follows, and people who have been taught to gain fulfilment in commercial directions are inexplicably treated as second-class citizens.

There are those who claim that civilisations just die of old age. Many have certainly done so and a large body of opinion would have it that such an ailment is natural within free societies. Decay, whether the cause be social or economic, is easy to pass off as inevitable. Religions are often the final consolation but history shows that these also die insidiously from internal crises of their own when social designs collapse under the weight of expanding bureaucracies and the masses feel abused.

It is regrettably true that life for the nameless millions, living and dead, has all too often been wretched, short and brutal. Few societies have secured peace or stability for more than one or two centuries. Throughout history mankind's masses have derived pitifully little from their existence, and drink, drugs, ritual and crime have understandably been favoured consolations. By promising improvements it is easy to demonstrate the superiority of a particular political system or the glorification of a race or elite class at the expense of others. There are those who believe that the human race exists just for the benefit of a select few, but fortunately this contempt for labour is a particularly precarious philosophy in the modern world where radio and television have made exact comparisons easy for all.

Many claim that a new type of structure is required. It is widely believed that mankind must look beyond democratic conservatism and socialism, with their unquestionable materialistic conception of life, and endeavour to find something different. Others believe that the problem lies elsewhere. The fact that the majority of Western civilisation is now divorced from the Church is often considered to have some relevance. Class consciousness and other social differences are also quite justifiably blamed.

En masse, people tend to look for salvation through a saviour in the form of a president or even a dictator. Unfortunately such leaders form only a very small proportion of humanity and it is difficult to see how radical improvement can come from such tiny numbers. The sum of human knowledge is quite obviously now so great that a single intelligence cannot even begin to contain the whole of it. The collective might of human knowingness is possibly not now capable of comprehending the multitudinous forces that interrelate to move society into cyclic phases of prosperity and depression.

The philosophy of politicians does not help. They believe that change is worthwhile only providing it is popular and can be realised in the comparatively short life of a particular government. There can surely be no such easy solutions to the complicated problems facing mankind. In a world where real distinctions are being deliberately eroded, or false ones invented, where reason is derided and the most fundamental principles of civilisation assassinated at an accelerating rate, real authority is required nationally and internationally. An impartial voice of sanity is desperately needed to preserve liberty, peace and justice. There is an urgent need to quash the lunatic revolutionary and racist unrest that threatens mass destruction. Religious beliefs, politics and nationalities are fought over hysterically, and need, greed and ego are undoubtedly a cause. Of course, the languages of words and numbers are frighteningly inconclusive, but the lack of tolerance in the world is difficult to explain.

Three and a half billion years ago organic life emerged on Earth. For this to be seen as other than just one event in one solar system of one galaxy of this one universe would be unjustifiably arrogant. However, untold numbers of different life forms followed the torturous journey, from corals to vertebrates, reptiles and birds, before mammals evolved. Mankind's history, if related to that of all other species, is not even comparable to the blink of an eye. Archaeologically, its ancestry is only traceable for thirty to sixty-five thousand years, and its social structures did not begin to form from the simple and illiterate farming communities of the early Neolithic Age until 5000 BC. Before this time it had lacked the concentrated power imaginatively to co-ordinate itself intellectually into hierarchical communities, and with division of labour into specialised crafts and professions.

The flood plains of the three great river valleys of the Tigris, the Euphrates and the Indus, as well as the Nile, then proved to be the right setting for its cultural growth. Unlimited water and immensely fertile soil, that could be maintained indefinitely by spreading fresh silt during the annual floods, provided a basis from which civilization and human imagination were able to expand. Minds then became emotionally charged with creative possibilities that extended far beyond the fulfilment of basic needs. Grand temples for prayer and palaces for earthly representatives quickly followed. Regular payments for the upkeep of these developments were then structured to aid maturing societies.

Cause and effect related to need and greed modelled the growth of the three valley civilisations, where social orders were formed to support necessary hierarchies. The machinery of urbanisation and city life, as well as many ritualistic behaviour patterns with their resulting demands, were soon accepted into the way of life. The framework then became established for the cultural development of modern man, whose intellect was ideally suited to such an adventure.

Evidence suggests that mankind and dolphins are the most likely species to become completely cosmically aware on this planet. So, should the aforementioned hiccup in mankind's journey to stardom really be looked on as a serious possibility? Astronomical observations have certainly indicated that a substantial asteroid is on a highly likely collision course with this planet. It need not cause immediate concern for it has a long way to travel and will take many generations to reach or miss Earth. But it does suggest that cosmic law could possibly have intervened. It seems obvious that mankind might not be conforming to required behaviour patterns.

This thought arose during the historic journey through the biological species of this planet. Mankind's forceful climb to its present elevated state shows a considerable amount of war paint. Whereas many less advanced species kill only to survive, the intellect of mankind persuades it to pursue directions that offer unnecessary fighting.

Throughout history, it is most noticeable that individual communities have found it important to protect those characteristics with which they associate themselves, whether they are racial, religious, political or economic. Groups of people, tribes and nations with these aspects in common tend to unite. They appear to do so in order to muster sufficient strength to overthrow, eliminate or subjugate those considered to be different.

Slavery of humans by humans to provide cheap muscle power has also been widely employed. It is possibly fostered by the deeply ingrained belief that superior races exist, so others should lead inferior lives and, on occasions, be shamefully exploited.

It would be ethnically convenient if mankind could argue that the classic form of slavery, the total ownership of a man or woman by another, ended with the earlier civilisations. Regrettably this is not the case for, although chattel slavery, the bondage of people having no human rights, is not now practised in the West, its demise is a comparatively recent development. The early Greek and Roman civilisations enslaved their own kind, but those sent to the Americas for nearly four hundred years to 1850 were exported from Africa. During the last two hundred years of this period twelve to fifteen million Africans were shipped across the Atlantic to provide labour for the mines, and to clear, develop and later work on the plantations of this new continent.

This enslavement of an economically disadvantaged people by another race, using the excuse that they are inferior, is an abuse that is not easily understood. Regrettably such activities leave a legacy for, although slavery of the black people of the Americas is now no longer practised, deep scars have been etched into the fabric of modern society as a result of this wholesale kidnapping.

In early civilisations there were vast numbers of slaves. Julius Caesar was purported to have sold 63,000 prisoners from Gaul as a result of a single battle. As much as half the population was in servitude in these times and the coming of Christianity did nothing to alleviate the situation. Aristotle's teaching on the natural order of things did in fact make serfdom respectable, and this did much to condition minds to the belief that inequality, in its varied forms, was acceptable. Changes came very gradually. Henry VIII freed the serfs on his estates during his reign and his daughter Elizabeth freed all those in villeinage in 1574. Other countries followed suit although forms of serfdom existed in France until the Revolution, and Russia did not conform until 1861. Up to that time sixty or more million Russian peasants had been owned by a quarter of a million nobles.

The seal of respectability was given to the trading of humans when, in the mid-fifteenth century, Pope Nicholas V authorised the bringing of heathens and other enemies of Christ dwelling south of Cape Bogador into perpetual slavery. But nothing compared with the extent of the slavery mentioned earlier that developed between Africa and the Americas within fifty years of the new continent being discovered. Thomas Jefferson, one of the great presidents of the United States, personally compared the imported Africans with whites and pronounced them inferior in reason and dull in imagination.

The blame for selling slaves in the Americas cannot entirely be placed at the feet of white men. The coastal kings and chieftains of Africa took an active part in the trading, to the extent that they herded prisoners into pens until ships arrived to take them away. The French, English, Portuguese, Dutch, Danes and Swedes were all very much involved. During the slaving period more than twelve million Africans were landed alive in the Americas. A further two million were lost during the ocean crossings and it is estimated that six to seven million were either slaughtered during the raiding of towns and villages to obtain captives or died in captivity while waiting for ships.

Inhumanities practised during the most active three hundred years of Atlantic slave trading seem to have known no limits. Slaves falling sick during the sea passage from Africa were simply tossed overboard, and women and men were whipped and humiliated constantly for the pleasure of the crew. Such recordings as that of a Captain Harding whipping culprits of an insurrection are commonplace. After doing so he made them eat the hearts and livers of others, and watched while a woman was hoisted by her thumbs and slashed to death with knives. It was normal for slaves to be branded like cattle with red-hot irons, and the women where constantly raped and brutalised in every imaginable way.

This quick portrait of mankind's behaviour is not the prettiest picture. Nor is it improved when it is seen that animals not blessed with mankind's consciousness do not needlessly fight their own kind in this way. After thousands of years of technological development mankind's aggressive instincts have not changed from those of the earliest settled communities. Wars continue to be concealed within various disguises of respectability and all relate to the failure to acknowledge what is transparently obvious to wild dogs. In order to survive they instinctively work as one. The advantage they gain is possibly difficult to quantify but it is easy to be persuaded that in this respect mankind is not as smart as these unaware animals. An important lesson has not been learnt. How to wear the acquired cloak of conscious awareness is the problem.

This discovery trail convincingly shows that there is no intuitively acceptable alternative to the way the cosmos works. However, mankind's consciousness, although as yet not fully explored or understood, does appear to create problems. It allows those so gifted to know that they belong to a world, whereas less advanced species are able to function only to the extent of their inherited genome intelligence. This is true whether they are ants, bees, cuckoos or any of the multitudinous species of the jungles and oceans of mankind's planetary home. In consequence these unaware species cannot envision death as mankind can, and it has been found to be a difficult realisation.

Innumerable religions have evolved which, to some degree, overcome the belief that mankind dies. An immaterial entity, distinguished from but temporarily coexisting with the body, is visualised and the religious beliefs promoted in support of such a concept are many and varied. Regrettably, vast numbers now argue theologically that their particular creed is the only animating and vital principle that is credited with truth. All offer the desired continuity of life beyond the period spent on planet Earth but sadly their minor differences cause massive social unrest and horrendous bloody wars. Incredibly these differences continue to divide society today.

Whether there is one valid religious truth and many unjusti-fiable ones is not an issue that this cosmic investigation is designed to debate. However, what is obvious is that these vary-ing religions propose nothing that cosmic reality does not guar-antee. Absolutely everything within universes is cosmic grain, which is eternal and never ceases to be consciously aware and all-knowing. This systemic bond allows inorganic and organic reality to be introduced to the cosmos in the form of the various structures of a universe. This grain that is in part mankind's physical self, its sight, intellect and consciousness, lives on after the demise of a universe and its organic life forms. Though it can conform to the behaviour patterns of inorganic atoms formed within a universe as well as to the independent adventures of its organic forms, this cosmic continuum is unquestionably immor-tal. The residual grain of all structures is the eternal omnipresent cosmos, a grain of total knowingness that functions consciously, ceaselessly and timelessly.

It can be argued that the motivation behind the unpleasant spectacle of mankind's behaviour is unnecessary, for it largely results from diverse religious beliefs. Surely these social barriers are completely artificial, for mankind's consciousness, which allows the realisation that physical bodies die, has exposed an eternal cosmic existence that includes periodic sojourns in a variety of activities within the universe.

The answer to the question, 'What is the cosmos and how does it work?' has not been made clear to mankind during its transition towards adulthood. Why this is so needs to be considered. Could a hiccup in planning have caused the cosmos not to have made clear to *Homo sapiens* that, fundamentally, cosmic grain, and therefore its knowingness and consciousness, are eternal? Because planet Earth's organic life is frighteningly temporary, to know its own systemic grain is everlasting could possibly have been very helpful to mankind. It can be argued that this knowledge might have resulted in less aggressiveness. However, it would seem to be obvious that the design of cosmic behaviour makes ascending species look for cosmic truth themselves.

A final word about orthodox religious faiths is possibly needed, even though it is outside the scope of this cosmic exploration. Vast numbers believe mankind's soul to be the disembodied spirit of a dead human being – a ghost. Cosmic grain is different, but intuitively my view is that the dissimilarities are not difficult to reconcile.

Now would seem to be the right time to discuss cosmic lore. Concepts such as right and wrong, good and evil need to be considered. Cosmic programming might conceivably respond positively to one perfect picture or many acceptable ones, all others being condemned simply because the required nuances of intuitive discourse are not there. Its lore must dictate the most practical way for its activity to take place and be consolidated within its structured omnipresence. Human behaviour, exposed earlier in the brief history of mankind, could possibly be judged as unacceptable to cosmic routines, which have to ensure that its cyclical behaviour patterns function timelessly.

Quite obviously, lore at cosmic level is difficult even to attempt to understand. At the human level it can mean an accumulation of facts, traditions and beliefs, all of which relate to each other. At the cosmic level it is a bit like evaluating good and evil; it becomes necessary to ask, 'Who's good and evil?' An archaic definition of *lore* might be of help: 'Anything taught or learnt.' And that is where it should be left simply because this discovery trail insists that all knowingness derives from the cosmos. Lore at cosmic level must be anything that is acceptable to its knowingness. It is unquestionably unchallengeable. How can it be otherwise? The cosmos is a timeless omnipresence without a beginning or end. Its lore cannot possibly be questioned, which is something that species *Homo sapiens* needs to respect.

Cosmic control has been mentioned and it has surely been seen that at times there is no alternative. Perhaps it would be necessary on planet Earth were mankind unable to pass 'suitability rules' relating to its total conscious awareness. It is conceivable that human aggressiveness, greed and selfishness could hinder the development of the delicate nuances of intuitive intercommunication required. Cementing the bonds so essential to absolute cosmic awareness might not then be possible and the dolphin alternative could become a 'cosmic must'.

Why should such severe action be necessary in this eternally evolving cosmos? The answer, if that is cosmic lore, surely must be that it is the way the cosmos works. It would be surprising if programmed tools that protect its cyclical objectives were not locked into the intellect of this omnipresence. Perhaps a realisation of its existence, with a clear understanding that there is no alternative to its laws, might attract mankind's respect. What is very obvious is that it is difficult to envisage any other eternal cosmos.

So: species *Homo sapiens* is built from cosmic grain, which is *mind*. This fact has surely been conclusively established. The grain of the cosmos is immortal, and its thought patterns instantaneously respond anywhere and everywhere. This grain is all mankind is, and in consequence no serious arguments can be presented to counter the view that, for brief spells, varying amounts of cosmic grain are locked into existing organic and inorganic structures. The grain's consciousness offers 'mature' species intuitive mental pictures that quite obviously are not yet fully comprehensible to mankind at its evolutionary level of awareness. But a great deal has yet to be explored and discussed on this subject. That mankind is not as yet dressed for the responsibilities of cosmic brotherhood is very obvious. The grain of its organic forms does not as yet react to the full extent of cosmic enlightenment, but its children or those of future generations can only reach this goal providing the route remains open.

While giving consideration to the sober issue of mankind's possible dilemma a far more uplifting intuitive thought emerged. These can pop up from the depths of the unconscious self like air bubbles released from the sea. Without warning or preamble they just emerge, and this one suggested that a totally enlightened species should be able to create organisms of its own. That such life forms could possess total knowingness and be designed to use it to the best advantage is an intriguing thought, but regrettably the manner and form of the model were not forthcoming, and the bubble disintegrated in a cloud of sparkling raindrops. Only mundane thoughts about the journey towards complete consciousness by dolphins and mankind remained.

It has been excruciatingly painful coming to terms with the collective behaviour of *Homo sapiens*.

Nelson is supposed to have written:

> *I wish I loved the human race*
> *I wish I loved its silly face*
> *I wish I liked the way it walks*
> *I wish I liked the way it talks*
> *And when I'm introduced to one*
> *I wish I thought 'What jolly fun!'*

15

Gossamer Thoughts

WHEN SETTING OUT on this journey one of my many intentions was that, at its end, it would be possible to portray the way in which the cosmos had evolved. At that time there was little understanding of its eternal omnipresence. It was a great unknown, appearing hazily on the horizon like a distant mountain range. Having travelled a little further, the view has changed. It is now recognised that the task would have been futile, for the discovery has been made that the cosmos itself does not evolve: it simply exists. Universes are, of course, different; they evolve, as do their galactic structures and the many and varied forms of organic life that are contained within them. As species develop within the countless solar systems of the cosmos, so do their genes that are the store of knowledge needed for their life patterns. They change into more sophisticated forms. But despite the great complexity and variety of this process and its capacity ingeniously to adapt to ever-changing circumstances, it has to be accepted that the behaviour of unaware organic life is in essence determined.

This key difference between consciously unaware species and mankind becomes very obvious when studying the former in their natural habitat. African safari parks are the perfect settings. Lions, leopards, cheetahs and elephants live with zebras, giraffes, antelopes and a whole range of smaller creatures, though often under the scrutiny of that ever-curious species, *Homo sapiens*. Given that they are quite obviously comfortably at home in such a setting, their apparent lack of interest in visiting tourists is quite extraordinary. It is particularly puzzling when they are so close to each other as to be within touching distance. However, the explanation for this apparently odd behaviour is quite simple: they lack the cognition associated with visual perception that consciousness allows mankind to enjoy. Their perceived indifference is an illusion: it is simply that they are not aware. This is not to suggest that they have an inability to sense and see their surroundings, but just that, at these particular times, their genes are not causing them to feel emotions such as hunger, sexual desire or fear. They are programmed to respond to specific stimuli and nothing more, having no appreciation of self and the world around them, so their actions are misunderstood. They are simply responding to inbuilt genome instructions which relate only to bodily needs and reproduction.

It is accepted that a great deal of human activity is genetically controlled in exactly the same way as in other animal species, and in consequence they have many similarities. However, when enlightenment from cosmic grain was acquired, many extraordinary differences were introduced. A whole additional dimension to their activity became available. Crucially, this consciously aware species was able to examine its reason for existence, and enquire into the nature of the surrounding universe. Sadly, the potential for intellectual expansion, rationality and freedom of enquiry have not always been fully realised. Restrictions have been allowed to take hold that have hampered the inquisitive spirit.

An example is the way in which fundamental enquires have

been distorted by the somewhat obsessive desire of the cosmological world to find a single law for everything. It has been its
Holy Grail for more than a century. Einstein and his associates
worked incessantly to find this mathematical consistency and
the approach of others has led to some very unhelpful and superficial views of the nature of the universe. Mankind has been
regarded merely as an anomaly and it has been suggested that
the universe and everything within it happened for no reason
whatsoever.

One of the rationales behind the preoccupation with 'unified
law' is presumably because it ties up with the idea of a big bang
beginning of the universe that resulted from a 'fluctuation' or
an 'irregular variation'. The problem is that so much has happened since then. Atoms, molecules, stars, planets, living cells
and life forms have developed in great complexity, but always in
conformity to law. In particular, consciously aware life forms are
able to reflect back a wealth of qualities such as beauty, harmony,
justice, providence and generosity. Such things make the mathematical approach seem totally inadequate. In consequence, this
enquiry does not accept the conventional scientific explanations
of this universe.

Einstein's famous 'spooky-action at a distance' observation
concerning atomic behaviour is possibly the right point at which
to begin an analysis of scientific inadequacies. It will be remembered that he made this remark when deeply concerned about
the implications for his own world view of a proposition made
by other quantum physicists. It was to the effect that, in certain
circumstances, measurements carried out on one atom directly
affect others, even though they are separated by large distances.
This seemed to imply a connectedness between atoms that was
even more fundamental than the accepted laws of cause and
effect. The old view was of a world of fundamentally separate

objects interacting by signals travelling at the speed of light. But these remarkable discoveries in quantum mechanics, which were subsequently verified by experiment, indicated an under-lying connectedness with the whole, a rather mysterious whole but one that intuition has revealed as cosmic grain, the intellect of the cosmos, an eternal omnipresence.

A more realistic model for this cosmic grain that brings it within the scope of our experience is simply *mind*. The cosmos revealed by this discovery trail can be envisioned not as a void or emptiness but as *thought* that is consciously aware and all-seeing. Atoms and organic cells are created from its grain, as is absolutely everything else. There is an old saying, 'As without, so within' – as in the individual, so in the universe. The two aspects are complementary and we can use our understanding of one to help us understand the other.

Thoughts are 'shadowy gossamers'; they are different from seeing, feeling and smelling. Their origin is the consciously aware cosmos which itself is beyond the distinction of inner and outer. Its immanent responses are its knowingness, and this is shared simply because everything is fashioned from the eternal grain of cosmic reality. Mankind, an evolving, consciously aware organic species, is learning to relate to this knowingness and the receptors of its brains have begun to evolve a connectivity that is totally different from that enjoyed by creatures limited only to the programmed data of genes.

It is surely not outrageous to describe the visions of cosmic knowingness as sculptured images, and it is quite exciting to begin to recognise and appreciate the concept of these intangible gossamers. Poetic nuances somehow explain these 'thoughts within', possibly because they are the language of the cosmos. Brains of consciously aware species progressively relate to its visualisations. Our preoccupation with the outer world and rationally evolved means of communication does, however, create difficulties when trying to communicate these visions. Words and symbols, though sophisticated when compared with the chatter of unaware species, do not relate to the intuitive discourse of cosmic adulthood. This is unfortunate because we use and apply these gossamers every day of our lives: they are what make us different from unaware animals. It is just that we tend not to appreciate fully the fact that our species is both genetically inspired and progressively consciously aware. We see and feel the difference, but obviously cannot as yet advantageously use the thought patterns of cosmic adulthood to intercommunicate between ourselves.

The communication methods of birds, lions, tigers and other animals convey genetic thoughts and intentions very coarsely. We do, of course, understand this. However, the subtleties that mankind has since introduced into language are huge, and obviously result from visualisations of cosmic grain released when awareness emerged.

It is regrettable that there is no alternative to passing on the gossamer thoughts resulting from our awareness other than by the spoken or written word. Nevertheless, thinking intuitively should be encouraged, for these thought patterns are those of 'ultimate reality'. However, conveying these thoughts accurately in words to others will continue to be difficult despite mankind's sophisticated methods of intercommunication.

There are those who say thoughts are sequential. They pass through time and space, and the words are heard in the mind. Intuitions then seem to be a kind of standing still, when a whole number of things are recognised in their relationship to one another – a train of thought, all at once, without a sequence. Others unquestionably think only in words, yet my thoughts are diaphanous. Their insubstantial gossamers envelope me and there is not a word in sight. Why thoughts are portrayed so differently is puzzling, and it seems important to encourage research on the matter.

Discoveries of modern physics have revealed that there is not anything that can be regarded as a material substance from which atoms are made. There is not really anything there but form and pattern, and motion governed by law. Recognition of this fact helps: there is nothing other than this 'grainy something', systemically structured into all atom activity. The quantum world's 'action at a distance' that so worried Einstein results from the fact that all these tiny structures emanate from the same source: their grain, which is mind, thought and intellect. It is no different from the source of the intuitive 'pictures within' that we appreciate so much, and its behaviour is instantaneous. In complex ways this inner intelligence ensures order and balance everywhere. It is not at all spooky but entirely natural, and atoms that are created by and from this grain make universes work the way they do.

There is no logical reason why a clearer understanding of cosmic intellect should not overcome the stone wall of incomprehension that the cosmological world has found so frustrating. Cosmic grain is that from which universes are built and through which they float as they expand outwards like gossamer balloon skins. Omnipresence dictates that the intelligence of the cosmos is not measurable. However, intuitions suggest that controls related to its stability exist. Its eternal omnipresence can never be other than secure. Its law cannot possibly allow serious imbalances within its structure that might endanger the whole or create situations in the large or the small that might do so. Its instantaneous thought patterns ideally relate to this level of omnipresent control, which would be so whether such instabilities are manifested where a universe is emerging, where black-hole cleansing is taking place, or where an advanced animal species is endangering the natural life of a planet.

Tsunami disasters, land-based earthquakes, hurricanes and tornadoes could undoubtedly be cosmic warnings that species are acting unwisely. This is so even though it has to be accepted that it is extremely difficult to relate such occurrences directly to geographical areas where objectionable behaviour is actually occurring. That these warnings are meant to indicate that there is a need for consciously aware species to get their act together and respect cosmic laws is implied by the fact that unaware species seem to be instinctively forewarned of tsunami-type occurrences. Disapproval at cosmic level is directed towards aware species and, in my opinion, these laws are not consciously administered: they are provided in nature's way, within cosmic grain's thought structures which relate to organisation. Global warming and excessive melting of the polar ice cap are symptomatic. The systems of the cosmos ensure that the structures of its universes work in accordance with its laws, and species evolve to ultimate adulthood when these laws are respected.

Seeing, thought and conscious awareness continue to be the most tantalising of all cosmic phenomena. As has been repeatedly stated, genes control the growth, regulation and involuntary behaviour of all animal life from conception to death. The extraordinarily sophisticated level of gene programming that is necessary just to correlate the sensory activity of seeing to the consequent motor response of an organism is but one minuscule example of its governance. Mankind does, of course, share this genome programming with unaware species, but in its case there is the crucial additional factor of direct access to consciousness that has become available, and evolutionary progression has enabled it to begin to mature.

143

Probably the only other aware models on Earth are dolphins. Unfortunately the images gained when endeavouring to study their consciousness and relate it to our own are easily washed away before useful conclusions can emerge. It then becomes difficult to relate these to emerging concepts, and half-formed shadowy structures remain incomplete. In consequence, this inner aspect of the world continues to remain the most difficult problem of all to rationalise.

One thing that has become clear is that the intuitive nuances of thought derive directly from cosmic grain. This systemic 'cement' of everything communicates in an intuitive language of feelings or pictures within. They are always everywhere, but only when organisms acquire consciousness can their brains be capable of giving meaning to its pervasive gossamer patterns. In the outer world the grain of the cosmos also has to be accepted as that from which everything emerges. The atoms of its omnipresence are undoubtedly its creative tools within universes. They fabricate inorganic structures as well as the varied species of organic life, with their inherent capacity to respond to external conditions and in higher forms display remarkable feats of skill in acquiring food and caring for offspring. What remains the key understanding, and the one we need to be continually reminded of, is that, ultimately, absolutely everything, whether it is viewed as life, substance, force fields or thought, derives from cosmic grain.

So, are these considerations and conclusions about thoughts and consciousness of any use, other than academically? Are they even enlightening concerning mankind's true relationship to the cosmos? From the point of view of the mind, the truths established during this journey of discovery are undoubtedly of considerable importance. Feelings that give rise to the belief that conscious awareness is part of an everlasting cosmic story take time to mature. A milestone discovery was undoubtedly that brains do not think. That all knowingness is a phenomenon of the eternal omnipresent grain of the cosmos, whether genetic or otherwise, will need to be digested with considerable care. However, the cosmic story still has some distance to travel. This mind adventure is formidable.

Thoughts are not words; they are pictures within. How can they possibly be conveyed accurately to others?

16

A Sex Act Model

THE COSMOS has been shown to be an omnipresence of instantaneous knowingness: this has now been established as a firm base for further explorations. What is of particular interest is how the cosmos manifests its intelligence through living creatures. In this regard consideration has already been given to the way in which an immense amount of data is contained within every one of the billions of microscopic living cells that co-operate to build complex multi-cell life. Yet this is just one aspect of cosmic grain at work.

It is known that all the cells of any particular animal species contain identical microscopic genomes. These chromosomal tapes work ceaselessly during the life of each and every organism. According to their pre-programmed instructions the cellular structures are built, the metabolic processes maintained, and their many functions regulated through sickness and health. In fact there is a view, a rather extreme one, that all unaware animals 'are their genes', that their selfhood is enshrined in these molecules, and that the rest of the cell is the means of maintaining and reproducing them. The genes not only regulate body functions, they have also been shown to adjust strategies, accommodate new circumstances and, with great precision, orchestrate many other functions related to the outer world.

These in-built knowledge banks of cells are the repositories for information for all this co-ordinated activity, and the way it is encoded at the molecular level stretches the concept of scale to extremes. Genes are so extraordinary that, although it is possible to accept the levels to which their pre-programmed data extend, a real appreciation of this phenomenon defies comprehension.

For aware species there is another source of knowingness. In addition to the store of their inherited genes they 'tap into' cosmic intelligence directly. This provides them with a conscious appreciation of self, and these attributes enable human beings to perform in the many ways that have been discussed. Because the cosmos is the source of consciousness, it has an awareness of all these experiences.

Intuitions have indicated that a cosmic aim for organic life, when it becomes sufficiently advanced, is to open the gateway to total enlightenment. Mankind appears to have been subjected to the preliminary stages of this process. The popular view, that awareness results from the evolution of the brain and that some-how it began to manufacture knowledge and use it to invent more of the same, can surely now be looked upon as purely egoistical. This much-prized organ is created during morphogenesis together with other organs and parts of the body. All are genetically contrived orchestrations, as is the data on how the whole living organism should be shaped and made to respond to everyday life. It is well known that genome instructions supply all the necessary information. The way young birds are able to build nests and migrate huge distances without parental instruction has shown that such activities result from 'seeing/thinking' genome programming, which is the only intelligence that unaware species can access.

It is obvious that the design of every individual cell of an organism is extraordinarily complex and ingenious. But suggestions that brain cells have an additional ability that allows their cognisance to extend beyond the knowingness of other cells are completely unfounded. All cells are adapted to perform in special ways during morphogenesis. Their form is changed when they are selected for jobs within a liver, heart or eye, but none are 'extra special'. The design of the cells of each organ, including those of the brain, is such that their genome data can be considered for all practical purposes to be identical. No evidence whatsoever exists to support the popular view that brain cells manufacture thoughts. They just do their particular job of co-ordinating sensory data and movement.

These new ways of looking at thinking do not have to be seen as frightening. Although they imply a radical reassessment of the brain's function, the previous theories have never been secure. No one has the slightest idea how the brain is supposed to have been capable of performing its accredited functions. It is now proposed that, though cells cannot think, they do dispense knowledge. Their genetic code is complex and its programming extraordinarily detailed. The chemical machinery that makes this immense flow of information possible is remarkable in the extreme. To achieve what they do, cells manufacture proteins, combining amino acids at rates of as many as fifteen million per chromosome a minute. However, it is not always possible for information to be delivered internally via chemical messengers. The picture needs to be extended.

All information has to be signalled. Mankind communicates in spoken languages and has designed alphabets so that words can be written down. The body's chemical messengers are another such language. Many of them are quite simple proteins. They travel through the body instructing cells to respond in ways that cause the whole organism to behave in an orchestrated way as a complex animal or plant. But it has now been seen that it is not just these genetically regulated transmissions of 'protein shapes' that cause aware species to relate to their world in the complex ways they do. Their responses to outside events demand thought patterns derived from an interaction with the cosmos as well as the 'nuts and bolts' language of species genomes. A more cogent explanation of *thought* is obviously necessary, for that of both genes and cosmic grain must be closely related. It would be extremely foolish to put them into completely different boxes. The example chosen to illustrate and clarify this is the human sexual experience.

It is well understood that the familiar and unmistakable physiological changes required to perform sexual acts are responses to complex and diverse stimuli ranging from physical closeness to sight, smell and, of course, desire. This pre-sexual arousal prepares those involved both physically and mentally for the act itself. Unconscious genome intercellular chemical signalling is involved – indeed for unaware species there is no other arousal: they rely completely on this inner language to excite responses. A reaction excited by a photograph, a letter or a telephone conversation demands a level of realisation that can only be associated with the intuitive language of the cosmos, its 'pictures within'. These are the nuances of expression and feeling associated with conscious awareness and they allow those so endowed purposefully to examine their own inner desires before the sexual act. These anticipatory intuitions derive from cosmic grain. The feelings are experienced throughout the body simply because the grain of the cosmos is everywhere within us.

What exactly is this enlightenment that is dispensed to consciously aware mankind? What are these intuitively excited pictures within? Are they completely free from genetic control? In the context of sexual excitement it can only be said that the difference is that between chalk and cheese. At best, the males of species that are not consciously aware 'scent' the genetically programmed need of the females and perform the sexual act as and when allowed. It is a procedure that is often performed quickly, and with no more thought required than for eating. Engineered by inherited genetic programming, the language of chemical messaging within serves its purpose resolutely.

The behaviour of men and women is quite different. Their sexual activity is carried out with gay abandon, usually by mutual consent. Importantly, it is enjoyed as an experience in its own right as well as a consciously planned route to produce offspring. The pleasure is quite often fantasised over to a considerable degree and very few would argue that it is not a beautiful adventure that should, like good wine and music, be enjoyed to the full. The nuts-and-bolts, genome-inspired inner responses play their physical role in this behaviour sequence, but the poetic nuances and comprehensive emotion within derive from the intuitive mental forces of cosmic grain's intellection.

The limitless grains of the cosmos should be seen as infinitely varying energy shapes that continuously and instantly interrelate everywhere. This grain is the fabric of atoms, which are the building blocks of everything within universes. It also fulfils an additional need in organic life, allowing advanced species to appreciate the holographic effect of thought.

It must be remembered that the totality of intellect or thought is, effectively, everywhere within every atom. This is not only because scale is not an issue, it is because thought is an instantaneous transmission. There is absolutely no time-lag. Consequently, all knowingness, absolutely everything, is always in every possible place. It is regrettable that, as yet, it is not known how fully to access, digest, savour and use it, but it is hoped that these arguments have been sufficiently convincing to show that thought is cosmic in origin. Organic genes can be regarded as an extension of this phenomenon: they are selected knowingness designed into the genomes of organic cells. Everything is saturated by this knowingness, the 'ether' of Newton's era. It was he who insisted that its responses are instantaneous. And it has been agreed that its laws are cosmic laws: there are no others. Its gravitational glue holds universes together and its intellection gives regulation to cosmic behaviour through the activity of thought. At opportune times it becomes possible for cosmic grain deliberately to react in specific ways, which can be looked on as talking to itself in its own pictorial language. Truths of the cosmos are, of course, introduced to organic species in this same way after they become consciously aware.

The widely publicised out-of-body phenomenon could possibly relate to conscious awareness. When it is experienced, men and women appear to see their own bodily form from outside in the way that 'cosmic seeing' has been described. Eyes are not needed to see the body which they recognise as their own, but it is clearly observed 'floating' in its own space. Recently Swiss doctors have reported that they can repeatedly trigger such 'images' by placing electrodes on a spot in the right cortex of the brain named the angular gyrus. Patients treated in this way convincingly describe looking down on their own bodies and the medical staff in the operating theatre. It has been reported that such occurrences are quite spectacular.

The angular gyrus obviously includes a mechanism that normally restricts seeing. Cosmic visualisation appears to have acknowledged that organic brains need to be so designed that seeing the world in the way that it sees is prevented. As discussed earlier, the orientation of the eyes of organic species

151

allows them to see only in limited directions. They would not otherwise survive. The gift of seeing would be disastrously confusing were organisms able to see within universes in an all-encompassing cosmic way.

Of great importance is the fact that, when using their electrodes, the Swiss doctors appear unintentionally to have unlocked cosmic safety controls within the brain. The patient has then enjoyed a degree of 'out-of-body' seeing, the cosmic way, completely divorced from the limitations imposed by the eyes of organic life. It is interesting to consider why such an easily adjustable control exists within the brain.

Perhaps another area in which the possibilities of access to cosmic grain have been exposed is in individuals with remarkable mental gifts, now known as savants. It is a phenomenon that came to the attention of the public through the Dustin Hofman film *Rain Man*, based on the life of Raymond Babbit, who was capable of extraordinary feats of memory. Others have outstanding artistic gifts, being able to draw with remarkable accuracy and fluency from a very early age, and with no training. In a number of cases these abilities have followed an accident or other event that affected brain function. A recently publicised example is that of the young Englishman Daniel Tammet. Following an epileptic-type seizure in early childhood, Daniel gained a remarkable ability with numbers. He can carry out complex calculations with great facility. The crucial point is that he does not work out the answer, but simply has it presented to his mind. The numbers for him are complex shapes that he 'sees' within. One of his public feats was to recite pi to 25,000 decimal places. He did this faultlessly by simply visualising the number. For him it is like a landscape. Interestingly, a neuroscientist tested him by presenting him with a bastardised version of pi with some of the numbers changed. Daniel was connected to skin sensors at the time and these monitored a very strong emotional response. This example illustrates very clearly how, although the brain plays a part in these activities, it cannot be the origin of the thoughts, but merely one element in their reception and transmission. Daniel's gifts are not purely numerical: he also has a remarkable gift for languages and can learn a new one in a week. Again this shows that the essence of the phenomenon is a direct and uncluttered access to intelligence. The intuitive basis of this was shown clearly when a camera team took him to a Las Vegas casino. Trying to count the cards at the crap table did not work, but when he simply used his intuition, and split an unlikely combination of three sevens, he was able to make a scoop win.

These examples of savants are rather extreme cases possibly intended to provide a lesson for us. Often the gift is accompanied by difficulties in other areas, such as autism. However, there are countless examples of creative genius and inventiveness in all walks of life – in the arts and in science. Flashes of inspiration come into people's minds in ways that are totally inexplicable simply in terms of brain and genetic function. They very strongly suggest access to the source of creative intelligence.

Questions:

• **Might cosmic reality, the grain of cosmic thought, design brains to be unlocked by advanced consciously-aware species?**
• **Could this be the planned route to a higher level of organic adulthood?**
• **Is it conceivable that mankind is expected to explore this approach towards expansion of mind?**

Colours, shapes and vogues

The colours and shapes within universes are stimulated electromagnetically by vast numbers of genetically co-ordinated species. Vogues of art, fashion and movement are different. Only consciously aware species appreciate their subtleties.

17

Consciousness Explained

A COMPLETE UNDERSTANDING of cosmic behaviour would undoubtedly have ensured a clearer delivery of this narrative. The truth within the unadulterated gossamers that express its intuitions would surely then have been conveyed with ease. It is an experience that would have been dearly loved. But the means of interpreting the all-pervading pictures within are not yet fully developed. Mankind's maturity is insufficient to be able to share the perceptions of cosmic grain through an exchange of essential pictures. This ideal remains to be fulfilled. But in this context it is worth remembering the claim of the eighteenth-century French mathematician Pierre Laplace, who suggested that a sufficiently informed intelligence would know how the universe works. The cosmos has to be the unitary whole he envisioned, the intellect of which is now seen to be its eternal and omnipresent grain. It is unfortunate that the only communicative tools available to describe it, apart from body language, are the multitudinous languages of mankind, and the graphic symbols of written words.

One of the most debated problems in science remains unsolved: how to define consciousness – which is taken to mean self-awareness. It is frustrating that an adequate explanation has not yet emerged during this journey. Not to find a down-to-earth explanation for its uniqueness would be extremely disappointing. Indeed, for me it would seem like failure. Nothing less than a full and complete understanding of the cosmos was sought; incomplete pictures do not command respect. There seems to be a need to look back, to review the ground covered so far, and then endeavour to make use of arguments and considerations not yet fully exploited.

Man's inadequacy in looking for an acceptable solution to this intriguing problem of consciousness within the brain has already been described. Given the design of this organ it is impossible to visualise how it could ever produce original thought. Such an idea has to be wishful thinking arising from arrogance. Whether a collective or individual experience, conscious awareness has characteristics that are sensual, emotional and intellectual. They allow the aesthetic appreciation of art, music and nature, as well as the rational application of knowledge and reasoning, and the simple and direct experience of the present moment. There could, of course, be a very easy explanation of why the phenomenon can be experienced, but it is possibly too undemanding to be acceptable. It is that awareness is not the product of anything else, and is not caused by anything else, but is just inherent in itself through cosmic grain, which does not have a cause but has always been aware. This simple solution, however, lacks intimacy: it overlooks the one aspect of consciousness that most people value – which is individuality. It also has to be accepted that, even if the brain is not the cause of our awareness, for most of the time conscious experience takes place in parallel with brain activity.

Perhaps a more satisfying picture can emerge from considering consciousness from the point of view of the cosmos: what does it mean for cosmic grain, rather than the individual, to *know*? It has been accepted that the grain of the cosmos promotes and orchestrates the controlled big bang eruptions of universes timelessly throughout its omnipresence. Afterwards, atoms fashioned from its grain shape the galaxies and solar systems that emerge. Combinations of complex molecules produce simple organisms and thereafter the programmed development of multi-cell species commences, their selective routines of organic life evolving under the regulation of the knowledge held in their genomes. It would seem that these same multi-cell activities are directed ultimately towards cosmically aware species in multitudinous universes. It has also been acknowledged that the routes followed would have to be conceptually similar. But evidence does suggest that cosmic awareness is not necessarily designed exclusively for mankind. The intellectual constitution of the cosmos dictates the routes to be travelled, the progression of species towards adulthood being fine-tuned to guide adaptability and suitability in directions that accord with the environments of individual planets.

It is known that there are genes within the genomes of planet Earth's *Homo sapiens* that have not changed since the very first nucleated cell emerged. This indicates that organic life is a single family and can be traced back directly to common antecedents. The route that has led to the conscious awareness displayed by mankind has been a torturous one and there is probably still a lot further to go. One way the record of its progress has been laid down is in its twenty-three pairs of chromosomal tapes, each with several thousand genes, the unfathomable complexity of which exists within every one of the billions of cells. If expressed in words this encoded text of mankind would be longer than eight hundred bibles, and reading it at the rate of one word a second for eight hours a day would take a century.

That so much encapsulated data can be stored in such a

confined space is difficult to comprehend. That it exists within the genes of every individual cell, to instruct unaware and aware animals how to grow, function and reproduce to continually increasing levels of sophistication, is quite amazing.

Collectively, genes are responsible for the evolution of all organic life within universes, which is where they are cosmically fashioned. Looked at in this way, the cosmos must know all that can be known about the creation and assembly of genomes. This is indeed substantiated by the fact that the consciousness we have acquired from cosmic grain enlightenment allows us intuitively to visualise its knowingness and envisage the growth and maintenance of creatures themselves. It has made possible our understanding that, because of its instantaneous transmissions of data, each and every grain of the cosmos has a feeling for the whole. We can also see that it is remarkable that we are then able to comprehend why so much genetic information can be designed into the genomes of all species. Aspects of this knowingness remain beyond our level of comprehension to any degree of exactitude, but it is still amazing that the validity of this kaleidoscope of total knowingness within each and every grain of the cosmos can be glimpsed.

It has been established that awareness occurs when brains become sufficiently complex to carry out the required switching activities that are needed to relate to the 'software' information sequences of cosmic grain. The all-encompassing cosmic mind need not therefore necessarily know what causes its own awareness, or that of consciously advanced organic species. It is not even required to be able intellectually to rationalise seeing or understand why activity within expanding universes can be seen. Its only need is to understand the evolutionary processes of inorganic substances and organic life to the degree that enables it purposefully to manipulate events along known guidelines. All other aspects of cosmic knowingness could have just existed eternally.

Put differently, it may only be possible for the cosmos to understand its own intellect to the degree that allows it to direct the fabrication of organic genomes. Similar reasoning suggests that an understanding of seeing may be an impossibility. Only the eyes, which focus the sight of organisms, need come within the range of cosmic 'must-knows', for they are built under genetic guidance. A rational understanding of conscious awareness may be as obscure to the cosmos as it is to mankind, which experiences it because cosmic planning allows its appreciation. This does, of course, only become possible when the required connectivity between the brain and the grain of cosmic intellect is established, which could come about through evolutionary changes within its structure. There is no absolute reason for the cosmos to know why it or mankind is consciously aware or how the phenomenon is caused.

Offering only an explanation for the behaviour of the cosmos, while arguing that there is a limit to enlightenment, does not appeal. To doubt the possibility of understanding the fabrication of sight and consciousness would be equally frustrating. It is surely better to trust intuition and investigate the matter further, for it seems far more real to believe that cosmic grain's cognisance extends to everything that is associated with its omnipresence. An eternal omnipresent cosmos may not be an easy concept to accept, but it is even more difficult to believe that such an intellect would not have an absolute understanding of its own abilities.

It will be remembered that the sex act model emphasised the fact that genetic data is specifically programmed to satisfy the demands of every aspect of the lives of unaware species, from conception to death, as well as a great deal of aware species' needs. That this data is signalled in genome language at very modest speeds along established, easily identified internal body highways has since been accepted, as has the fact that the data is diverse and detailed. The grain of the cosmos is quite different and, being fundamental to the atoms of all structures in a universe, it causes them to be perceptive. The 'spooky action at a distance' observed by Einstein is related for, though unknown to him, the responsiveness of the atoms results from the fact that they are structured from cosmic grain. This means that these building blocks of everything within universes have the potential to appreciate to some degree cosmic grain's intuitive understanding.

The consciousness of cosmic grain has been seen to be something quite different from genetic intelligence. It is an enlightenment and knowingness that is always everywhere and, as the 'processor-switching' sophistication of brains evolves, this current of intuitive understanding is 'switched on' to the atoms of sufficiently advanced organic species. Quite obviously there is a need to become more familiar with this grain, from which all atoms are built. It is the substance of the cosmos and, as such, it has to be recognised as the 'engine room' driving the behaviour of everything in a universe.

Before commencing, it is hoped that scene-setting will be forgiven. It seems important to describe the type of intuition that somehow epitomises cosmic grain at work. It did indeed inspire the first thoughts in this résumé related to consciousness.

It occurred during a televised programme featuring a world-renowned philharmonic orchestra. A twelve-year-old boy was introduced as having a special talent and, as the concerto commenced, the television cameras were cleverly focused on this solo violinist's upper body, hands and violin. The unmarked fingerboard of this most sophisticated and difficult-to-play instrument, and the movements performed to achieve the results required, were beautifully pictured. The hand positions, finger routines and string selections essential to the creation of the desired patterns of notes and feelings were beyond belief in their complexity. Yet this young man's eyes never once looked at his hand. He was completely immersed in the magical world of intuitive thought patterns that related to his extraordinary talent.

As mentioned earlier, the remarkable facility with which Daniel Tammet is able to deal with numbers – that are just presented to his mind as is often the case with language – does suggest that creative intelligence is out there somewhere. For a special few this advanced level of thinking is intuitive.

This exploration does of course acknowledge that it is the engine room of the cosmos that ignites the intuitions of this young musician. He sees, feels and applies the 'pictures within' that make possible such performances, whether physical, mental, or a balance of both.

The structure of cosmic grain is what Albert Michelson's very sensitive split-light interferometer measurements were unable to detect in 1887. Originally known as 'ether', Michelson convincingly showed its mass to be such that it does not relate to matter in any form that is known. Universes float through it without interference and spawn galaxies with unique planetary systems that produce organic structures of many and varied species. And there is now no lack of understanding that such substance exists. As reported earlier, the United States National Science Foundation has acknowledged that only five per cent of the mass and energy in mankind's universe is detectable. Ninety-five per cent is considered to be deeply mysterious dark matter and energy that cannot be measured. Even more recently, in the year 2003, Dimitris Nanopoulos, as mentioned earlier, confirmed these figures. This 'mysterious something' most probably relates to the cosmic grain of this résumé.

It has been recognised that all the complexities of the multitudinous universes of the cosmos are constructed from atoms that are designed by and built from cosmic grain. In consequence, universe structures must all-pervasively relate to it. Even the genes of organisms are of its substance and the 'gravitational grab' of this intellectual force holds everything within universes together, exactly as Isaac Newton claimed but could not conclusively demonstrate. A final model of everything should ensure that this 'engine room' is understood. Its setting will be a three-dimensional television screen with an animated picture of cosmic activity.

The tiny bits of cosmic grain are clearly seen everywhere, some serenely floating in their own space while others are designed into functioning atoms within expanding universes. All variants are locked into the molecular structures of galaxies, their planetary systems and atmospheres, mountains, volcanic events, oceans, rivers, animals and plants. Although light-years apart, individual universes clearly expand freely through the very grain from which they are derived. Big bang beginnings of new universes are observable as cosmic grains fuse themselves into the shapes of the 'bits' from which atoms are known to derive. It even seems possible that universes 'interleave', which to some degree explains the Hubble telescope discovery mentioned earlier. It will be remembered that galaxies were sighted at the big bang centre of planet Earth's universe, which does not conform to the standard theory that the centre of a universe is empty, being the point from which everything explodes outwards, like a balloon's expanding skin. However, the stars discovered by the Hubble telescope could be explained by the interleaving of universes.

The model's cosmic amalgam of universes interact to the degree that individuality is lost within co-ordinated patterns of thought, every one of which is always effectively everywhere. In consequence all intuitions are as one, waiting for focused attention, as is the case with any picture display. Every aspect of universe behaviour is portrayed within this all-encompassing grain of the cosmos, which also structures atoms, the building blocks of absolutely everything. This activity moves outwards freely as universes expand through the grain's static omnipresence. There is so much space, order and disciplined acceptance revealed by this model that it is tempting to believe that the static grain swaps and changes with that which is moving outwards, enmeshed within the atoms of individual moving and evolving structures. As was the picture of the young violinist's finger movements, it is a beautiful visualisation. The static grain, and that moving within the atoms of universes, ensures that total comprehension of everything that is knowable is pictured everywhere. In effect, three-dimensional intuitions of all possible thought are eternally within every speck or fleck of the cosmos. Reciprocity is instantaneous and there is no competitiveness whatsoever, neither are the values revealed related to good or evil. How can it be otherwise? Everything is a matter of fact.

The plausibility of such order, where chaos is not always preventable, can be appreciated to some small degree from a study of any computer word programme. When a single phrase within a document is requested for inspection, it is found and produced the instant the demand is made. Transferring the whole or part of a document to another distant computer is accomplished without noticeable delay. And computers are very slow; indeed there can be no real comparison with the instantaneous rapport of cosmic thought.

Genes obviously take their place in this model and their know-ingness can be observed as it is dispensed along specific high-ways within organisms, for purposes that have been discussed. Brains are observed competently co-ordinating and distributing genetic data within individual organisms to meet the mind-shattering unaware demands of every organism of each and every species within countless universes. The management of cosmic grain's enlightenment within species becomes obvious, as they are seen maturing to consciously aware levels. This phe-nomenon appears to be triggered by sophisticated switching devices within brains that protect the unaware from the devas-tating flow of cosmic grain's total realisations until a species reaches the necessary evolutionary level. As conscious adulthood matures, species are seen mentally intuiting with the picture-thought patterns of the cosmic grain in which they are immersed.

Previously discussed genes, some of which have not changed since the first single-cell organism emerged on planet Earth, are observable in the genomes of a vast range of species. One is reminded that it is possibly their responsibility to cause brains to 'switch on' to cosmic grain knowingness when other cause-and-effect evolutionary routines have progressed success-fully. An alternative possibility is that the necessary changes result from cosmic grain's own surveillance and action. Its knowingness would appreciate when species within its universes had evolved to levels where awareness was desirable. Regret-tably, viewing the model did not clarify which method applies. However, it is sufficient to know that conscious awareness becomes apparent at appropriate levels of evolution, which occur when brains are 'switched on' to the knowingness of cosmic grain.

The two entirely different ways in which knowingness is dispensed are clearly displayed. Genetic data within organisms is signalled in genome language along well-used highways. The complex individual needs of all organisms are then graphically revealed. Enlightenment received directly from cosmic grain is quite different. Its intuitive thought patterns are eternally and instantaneously everywhere, systemically structured into atoms that build mountains, oceans and organic life. Every possible sensation, vision and thought, the totality of environmental possibilities, the essence of organic life's attitudes, opinions and feelings are observable everywhere.

Cosmic grain invasively relates to atoms, the building blocks of living cells. So, when released from the exclusive bondage of genetic governance, multi-cell organisms are seen to be totally transparent to its intuitions. Awareness then becomes evident in the form experienced by the cosmos at levels that can be limited only by the degree to which atoms comply when living in accordance with the physiological characteristics of an organism. It is, however, impossible to be 'switched on' to the omnipresent knowingness of cosmic grain and its related awareness without attaining an appreciable degree of that which the cosmos enjoys completely and timelessly. Absolutely everything is an amalgamation of its grain, which predominates throughout the cosmos whether the blend is a butterfly, a human being or a mountain.

Differentiating between conscious thought and that emanating from the genomes of living cells is not easy, even though gene knowingness is to some degree robotic. Inspired from within, it is excited by such feelings as hunger, desire and fear, and localised sensations stimulate activity in areas such as the stomach, genitals and bowels, and a vast array of muscles. Mostly they specifically relate to areas of the body where activity is expected or demanded. Some is even arranged so that it gives a feeling of pleasure, anger or pain, which again stimulates necessary reactions. These genetic responses, that are excited by what is seen or craved, play an important part in the inner activity of genome management, and it has to be remembered that all signalling is conveyed along specific organic lines of communication at modest speeds.

In contrast, conscious thought is immediate and derives directly from the grain of the cosmos, which is explicit. Its experiences – thoughts if preferred – relate to atoms, and it could be argued that such stimulations are external. They are not the tummy-rumbling hungry, 'I want food' type of feeling. They are not even the gnawing-within-the-groin feeling related to the tantalising demands of suggestive intimates. They are 'flashes' in which pictures of thousands of sequentially related images are revealed in an instant within the atoms of the whole living form. Brain-processor activity then reduces it all to meaningful kaleidoscopic multi-dimensional intuitions that relate to the timescales of organic life.

Mankind's every bit relates all-pervasively to cosmic grain, and awareness results from this fact alone. It is not a question of just being able to look into a mirror and understand what is reflected. The manifestations of knowingness, vision and consciousness are direct cosmic stimulations, and each and every bit of an organism's inbuilt grain, and that of the cosmic whole, are required to be fully aware of the pictures, images and thoughts. The process relates to the cosmos and causes advanced species to be aware of themselves and their environment.

Both genetic and cosmic knowingness are essential before conscious awareness can be experienced by an organism. Mechanistically speaking, they are two halves of the same coin.

The sophistications of awareness result from an interaction with the thought patterns of their systemised cosmic grain. Subjected to a degree of imaginative visualisation, deriving from experiences such as watching the earlier-described finger movements of the twelve-year-old violinist, this grain activity can be likened to small gemstones with many polished facets reflecting all 'thought' ubiquitously. Every bit of every atom of every cell within consciously aware multi-cell organisms has an awareness of the whole. This is because these tiniest possible particles, these thought patterns, have instant access to the conglomerate whole. It is an activity that can only be conceptualised fleetingly, but at such times it is possible to appreciate that this all-embracing mind-body phenomenon cannot do other than manifest conscious awareness from within the atom structures of an organism by interacting with the complexities of the cosmic whole. It is a feeling, picture or thought that ostensibly includes everything multi-dimensionally.

The activity of seeing is shared by the cosmos with a whole range of species that do not enjoy conscious awareness at any level. Their seeing experience relates to the omnipresent grain of the cosmos but an important difference has already been comprehensibly discussed, when it was shown that the electromagnetic forces of universes excite sight. It then became intuitively obvious why the cosmic whole sees only within its universes. These forces are needed to excite cosmic grain to 'see'. Organisms perceive for similar reasons. The total experience of sight is stimulated and experienced within the grain of their every atom. The orientation of the eyes introduces the reflected electromagnetic rays of the outer world to the systematised grain of an organism's atoms. Conscious awareness is not a prerequisite. However, as discussed earlier, extraordinarily sophisticated genetic programming is necessary within organisms to correlate internal perceptions to external positions.

'Omnipresent seeing' does not require eyes and this has been discussed. An interesting related thought is that only the physical organs, the eyes, of the blind are defective. The systemically fashioned cosmic grain of their atoms is excited by the electromagnetic forces of universes, as is the case with those who are sighted. The blind should therefore see in the cosmic way within universes when their species evolve to the higher levels of consciousness and are able to see ubiquitously.

Early during these travels the suggestion was made that the genomes of species are possibly programmed to expand their own enlightenment towards conscious awareness. Because the early changes appear to happen quickly, it seems probable that cosmic grain's own integrity causes them, rather than a long evolutionary process. Whatever the case, genetic knowingness identifiably stored within the genomes of all cells must surely be calculable.

A dilemma possibly arises when endeavouring to picture the actual configuration of cosmic grain's own knowingness. The omnipresence of the cosmos is the problem, it implies unlimited knowingness and such a concept introduces thoughts that relate to 'omniscience'. During this journey it has been agreed that such views should remain open to debate. So a Pandora's Box of questions presents itself. Hopefully it will not be found to contain all the ills that plagued the original, for it needs to be opened.

The cosmos is, of course, eternal and ceaselessly consciously aware. Though periods are spent within the temporary atom structures of universes, continuity of its grain's existence is unquestionable. It has been acknowledged that it is invulnerable, but could it be claimed that its omnipresence demands that its knowingness must be unlimited?

It can surely be claimed that cosmic knowingness extends to whatever level is required to attend to the needs of universes, both intellectually and physically. However, it can also be categorically stated that the grain's design is undefined, so the amount of data contained within each cosmic grain, and the number that carry identical knowingness, remain debatable. Cosmic knowingness could therefore be computable, as is the case with gene knowingness. However, because of its omnipresence, its knowingness could be unlimited. Pandora's Box can therefore be quietly closed, for thoughts related to omniscience can continue to be debated.

These travels within the cosmos have now reached a point where it can be categorically stated that what is 'out there' within the ubiquitous cosmos cannot be understood in greater detail. It can also be said that the limits to the intuitive knowingness of species within universes can unquestionably be refined and extended beyond that of mankind. Our species could and should relate to the observed sensitivities – cosmic law if preferred – more advantageously, and make greater practical use of what is within the cosmos. Mankind's knowledge can surely be extended but, because we are not as yet cosmically adult, we are unable to evaluate to what degree.

Cyclotrons capable of generating particle energies at tens of millions of electron volts have been used to investigate matter. It is generated at a central source and accelerated spirally outwards in a plane at right angles to a fixed magnetic field by an alternating electric field. The objective is to break particles down into the smallest possible bits.

So, can the make-up of cosmic grain be explored further?

The answer to this question surely has to be 'No.' That there could be such an inherent cosmic weakness is inconceivable. This grain cannot be destructible or changeable. The decision made by America to release their first atomic bombs on Japan to end their lingering war shortly after the second war with Germany had ended is sufficient proof. At that time it was not known for certain whether atomic chain reactions could be contained. Only one universe was at risk had its foundation been in danger. The grain of cosmic reality has to be indestructible. No cause can possibly exist that might result in its dismissal or change its integrity. For this reason alone it is obvious that cosmic grain is not only omnipresent, it is an eternal presence that is irreducible.

Big bang beginnings of universes occur, as do black hole reductions of their worn-out debris. These compression exercises of the cosmos cause excesses of its grain to be locked into the purposefully excited big bangs of new universes and black hole reversals when undesired rubbish is reduced to freely floating cosmic grain. A balance is thereby maintained causing the knowingness and required transparency of the grain of the cosmos to be maintained.

Every intuition or 'picture within' is obtainable from every cosmic grain. But does this mean that each grain is cosmically explicit?

The answer is 'No.' The grains of cosmic omnipresence collectively constitute its enlightenment. It is intuitively readable from anywhere because inter-grain transmissions are instantaneous.

18

Awareness Responsibilities

THIS ADVENTUROUS JOURNEY has been an awesome experience. Many discoveries have been made and intuitive thought has undoubtedly helped establish truths related to the cosmos and its workings. However, not always being able to find a way of portraying these intuitions that does them justice has been somewhat frustrating. Because of this, it is believed that some of the additional insights that have been glimpsed and recorded in these concluding chapters will hopefully be of interest. The images that were envisioned offer considerable scope for lateral thought in particular directions.

It has been repeatedly felt during these travels that words are clumsy communicative tools. At times the significance of certain aspects of the journey have not been given the attention they needed simply because finding the right expressions to describe envisioned disclosures was impossibly difficult. These very disclosures have forcefully revealed that words can in no way take the place of the intuitive discourse that is the only language of the cosmos. This is because cosmic intuitions – its knowingness – are thoughts deep within mankind's organic shell, which need space and calm to reveal themselves. For mankind in general the difficulty in making this inner connection is compounded by the insatiable desire of the ego to indulge in and be driven by competitiveness. This leaves little room for true insight to emerge.

There seems to be a need for a suitably relaxed environment and company in which it would be possible to begin to communicate to a greater degree without such distraction. It is feared that very little headway will otherwise be made in addressing the many extremely urgent matters that face mankind. Surely the most important task now is to learn how to become 'tuned in' to cosmic reality and stop being distracted by trivial matters which make us 'fiddle while Rome burns'.

Cosmic planning of universes and their structures will not be put on hold just because we ignore them. Although the unitary whole is invulnerable, mankind itself is not indestructible – or indeed 'the only game in town'. This surely needs to be understood. An important realization that this discovery trail has uncovered is that mankind's behaviour alone will determine whether the species continues to be an important part of cosmic history.

There are so many issues that need to be faced up to. One of the most crucial is surely the need to engage in an effective way with that most fundamental of studies, the pursuit of wisdom. The Greek word *philosophy* literally means the 'love of wisdom', but in this age it seems to have been reduced to the level of intellectual wordplay and to have become divorced from the way in which people actually lead their lives. Looking around at the world, the apparent inability to find resolutions to the extraordinary differences in opinion on matters related to religion, the distribution of wealth, overpopulation, access to and preservation of scarce resources and the right to pollute planet Earth are frightening. Select communities too often consider that they can reign supreme, and the difference between what people, including politicians and other leaders, say and the way they act is difficult to justify.

Looking back in time beyond the feral state of our ancestors, the Rubicon, the point from which there could be no turning back, was unquestionably the time, whenever it was, when conscious awareness first became available to man. It was then that our forebears were able to realise that both they and planet Earth existed. This consciousness surely ought to have allowed aware species to begin to *feel* the intellect of the cosmic whole and relate to its intuitions that are all-knowing. The difficulty was that, through vanity, intelligence became associated with the individuals who received the thoughts and not with their true source. This is like believing that a radio programme is coming from within the receiver rather than being broadcast from a distant studio and only being received and translated into audible sounds by the set. Man's overwhelming desire to be 'the best' is possibly also very much to blame, for in this context a good brain is seen as a prized possession to be used for one's own benefit, and not as a precious gift to be cultivated for the sake of all.

Earlier in this cosmic exploration the sex act model was used to portray the dual effect of the programmed data of genes and the knowingness of cosmic grain. Although both are signalled to the brain, it has been recognised that the methods used are entirely different. The reactions of all species to the genetic knowingness of their cells have shown its programming sequences to be superb. From the moment of release from the womb newly born animals instinctively breathe as independent living entities and, when hungry, they know how and where to suckle the mother's milk. All this and much more is now known to result from genetic instructions ceaselessly dispensed from the mass of living cells within each and every organism.

In creatures with conscious awareness there is a another path to knowingness – direct access to the intellect of cosmic grain. For mankind one highly significant effect of this has been that the grunts and groans of unaware animal intercommunication have been refined into precise words and grammatical arrangements. When reviewed earlier it was accepted that the complex chattering of many unaware species suggests that the structures of their 'language' are stored within their genomes. The limitations of their systems of communication seem to result from their lack of awareness, which greatly limits the development of syntactic expertise. This limitation would also apply to symbolic logic. The intuitions of this journey of discovery suggest that systems of symbols representing quantities and relationships, i.e. mathematical logic, remains quiescent within the genomes of species until they are inspirationally aroused. Such arousal causes the flood of intuitive 'pictures within' experienced by people exposed to the modest levels of cosmic awareness that we now enjoy.

It has been clearly shown that thoughts originating genetically are quite different from the patterns that develop as a result of conscious awareness. Intuitive thoughts are those derived from cosmic reality and they offer a potential awareness of everything. The offspring of the consciously aware can objectively and subjectively enjoy and question everything.

As more sophisticated species evolve, their brains mature to levels at which they are able to process both genetic knowingness and insights received directly from cosmic grain. This distinction was demonstrated by the South African naturalist Eugène Marais. He describes how, during a drought in the Waterberg, he rescued a baby otter, which was dug out of its nest shortly after birth. It was taken some distance away and reared by a bitch with the rest of her litter. At the same time he took a baby baboon away from its home in the mountains and reared it on a bottle rather than its natural diet. After three years he returned both to their natural environment. Back at the Sterk River the otter, although it had never seen running water before, after a brief hesitation took the plunge, and within half an hour it had caught a crab and a large carp and devoured them. The baboon on the other hand was completely lost even though it was surrounded by its natural food. It had to be shown how to find insects by turning over stones and how to avoid being stung by scorpions. Marais had made a careful study of baboons. He recognized them as highly intelligent, adaptive creatures with the ability to learn: animals on the brink of conscious awareness and no longer exclusively under the limitations of genetic knowledge. It seems there is a price to pay for this newfound freedom: the young of each new generation have to go through the process of learning.

This maturation takes place throughout the life of each individual organism. In humans intuitive thoughts developing from cosmic grain are not readily available to the newly born even though their potential for awareness is considerable. It would not be wise for the cosmic enlightenment of the mature to be available to their offspring immediately. Indeed, problems relating to certain child prodigies could well be the result of some form of cosmic displacement in this direction.

As young brains develop, interconnections between the nerve cells appear to be built up as the child responds to the stimuli it comes into contact with. In time highly complex neural networks form, and the more stimulation that is received, the more they grow. This is a process that requires great care if it is to take place in a complete and healthy fashion. A considerable degree of discipline is required within families and communities to ensure that the appropriate stimuli are provided at suitable times and in the right measure. There are critical learning ages for particular stages of brain development and it is important that educational institutions play their part in exercising the required disciplines.

For the first five years children are very tender and should possibly be fed primarily on love and affection, to which they respond very readily. All they need to be taught they can receive simply through love and play.

From five to ten there is a great capacity to acquire simple knowledge and this should be made available. If children like something they will pick it up easily and copy anything coming to their senses. Anything new is full of interest, but this can be a distraction. Simple discipline is needed to provide order and regularity so that learning is efficient. The pitfalls of being unmethodical, loose and disorganized, and of being diverted into crude and base activities, need to be avoided.

From ten to sixteen there is a rise in mental energy and the intellect begins to develop. Children are able to learn to relate things and to find causes.

After sixteen the state of adulthood is reached but there is still much to learn. Responsible work begins and the young person has to develop as an independent being. It now needs advice rather than orders.

It is well known that speech-pattern learning must not be delayed into adulthood. There are times when learning passages by heart is quick and easy and good material picked up at the right stage can then be available whenever it is needed. There are times when words and grammar need to be linked, otherwise these essential arts become difficult to master. Quite obviously the consequences resulting from ignorance or lethargy in this direction can be disastrous. If they are not promoted at the right time they can never be completely corrected. Normal life then becomes difficult. The nuances of cosmic enlightenment and elementary genetic data related to language and other areas of learning need to be consolidated within carefully agreed timescales. Specialised group teaching has an important role in all education.

Although it is now known that wisdom is a cosmic phenomenon, it cannot be overemphasised that it is not just there to be 'plucked out of the sky'. Its successful attainment would require changes to be made to many teaching methods. These will be absolutely essential to this radical new vision of knowingness.

If there is no acknowledgment of how the cosmos really works, then responsibilities related to awareness may be attended to with little skill. Such ignorance does not interfere with the running of genome software simply because that is handled superbly below mankind's individual level of consciousness. To access cosmic wisdom will require detailed planning by neuroscientists, philosophers and psychologists, who will need to develop and promote the required understanding.

There are many directions in which research could be rewarding and yield positive and exciting results. Establishing the degrees of difference in the capacity of newly born individuals to store and recall cosmic data would be one obvious direction to follow. The effectiveness of individuals might not vary a great deal given that all react to genetic data with similar levels of competence. This is so despite the fact that a great deal of genetic knowingness is far more complex than much of what is received as a result of cosmic awareness. The notion that some individuals are gifted with superior intellect could have been very much exaggerated. Perhaps it is not that some individuals are inherently more intelligent, since there is only one real source of intelligence, but just that they have benefited from a better quality of input. This may well be the reason for the difference in knowingness. Mistaken reasoning in this direction could also result from the fact that better-informed minds are often found amongst the progeny of the wealthy. Early parental stimulation of children is an explanation, as is the fact that their education has the benefit of being very fully resourced. The assumption that the accumulation of wealth results from superior mental attributes that are passed on genetically is highly questionable.

An understanding of whence knowledge derives would help eradicate many deep-seated false beliefs in this area. It would show why finding the best ways to organise mental input is most rewarding, as is ensuring that appropriate teaching methods are offered to all. Providing educational material of the highest quality is important in this. Basic education should obviously not allow faulty input from uninformed or disruptive sources to contaminate the young, although there are no known ways of completely eradicating such input. The 'viruses' with which the 'mind of the internet' is cursed graphically illustrate the seriousness of this problem. With sufficient care and vigilance much could be done.

Creativity should certainly not be overlooked. The art of searching for thought at the cosmic level should be taught with patience and care. In this there is possibly a need for far greater use to be made of visual impressions. It should be remembered that they are the only language of the cosmos. Normal verbal methods of communication can be too coarse; they do not portray the desired nuances of a cosmos that cannot talk. Its intuitions, the pictures seen and felt within, need to be explained and encouraged. It does, however, need to be recognised that vast numbers of people still think only in words, the significance of which can so easily be wrongly interpreted and manipulated. 'Think in pictures and talk in words only if necessary,' is the best advice.

The way in which the intelligence of the cosmic mind can be connected to intuitively has been discussed at length. Its omnipresence is gossamer, an indeterminate fabric of total enlightenment. This resource of continuously available knowledge needs to be understood alongside that of the genomes of living cells since the process of thought relies on both sources. Each one plays a part in reflex actions and purposive behaviour patterns within. Human languages do only the postman's work, delivering data syntactically to other members of the species. Original thought eminates as pictures or feelings from cosmic grain. They are seen or felt by species that are aware. It is here that a human dilemma is exposed.

Organic genomes orchestrate morphogenesis, body functions, and many living needs, including those that excite organisms towards crude intercommunication. Feathered birds twitter, while land and sea species grunt and chatter. As well as alerting comrades to danger these basic languages play their part in the rituals associated with reproduction. When conscious awareness emerges, the intuitively inspired enlightenment that then becomes available needs to be debated and used constructively. Unfortunately many unnecessarily varied languages have developed and the complexity of these modifications of the gene-inspired grunts and squeaks of long ago have inhibited global discourse. One is tempted to condemn the cosmic mind for such an evolutionary hiccup, but this would be to misunderstand cosmic reality. Problems of this type are benchmark tests for aware species to solve and it is regrettable that there has been little progress in the direction of a single global language. Indeed, mankind has done very little to help itself since the advent of its conscious awareness, the greatest gift possible within the cosmos.

Reason suggests that it is unfortunate that cosmic 'pictures within' do not conclusively show that good and evil are illusory concepts. Earth tremors and the cataclysmic eruptions that occur within universes, galaxies and planets receive no such evaluation from mankind. They are looked on as evolution at work. So should it not all be seen as one and the same thing? It cannot be rationally argued that cosmic regulation should prevent little Willie from being bullied by bigger Tommy, or that it should stop a million ants from being slaughtered by larva erupting from an active volcano. If mankind cannot establish the necessary disciplines to survive on Earth, other cosmically inspired organic species will take its place. Success and failure are only in the eye of the beholder. It has been shown that planet Earth might well have the necessary time to evolve new species should mankind fall by the wayside. Species surely survive if ways are found in which to do so!

Awareness has made possible the realisation that mankind's activities could make planet Earth environmentally unsuitable for organic life. This journey of discovery acknowledges that cosmic law could be designed to 'kick into play' at such times planetary activity that would destroy mankind if it were dangerously failing to conform to cosmic safety laws. In that event, dolphins might take over, though it is suspected that the primary reason for their sea-based evolution is to insure against an unforeseeable natural land-based disaster. It would be most unfortunate if the combined efforts of Earth's many independent religious orders and political regimes were unable to ensure that the route travelled by man will not cause planet Earth to be mortally wounded. The dangers are glaringly obvious, and the design structures of cosmic behaviour unquestionably do everything possible to ensure that all is transparent to consciously aware planetary species. But it is often said that there are none so blind as those who do not wish to see and these words might well become the epitaph of Earth-bound *Homo sapiens*.

Winning surely causes jealousy, resentment, and anger.

- **Why is this pointless vanity so prevalent?**
- **Is popularity in any worthwhile form expected?**
- **From those who are made to feel inferior, or those who caused this feeling?**
- **Surely not!**

19

Inner Reflections

NOT PRACTISING what one preaches is one of the easiest of accomplishments, especially when philosophising. After such an exciting and privileged journey I hope that observations made now will not be looked on harshly. There is absolutely no doubt whatsoever that my need to learn from the discoveries made during my travels is at least as great as that of any reader.

Intuitions, whether philosophical or otherwise, are undoubtedly delightful experiences. It is difficult to see them other than as thoughts from within, and this journey has certainly convinced me that this is so. They always seem to pop up from nowhere. They just emerge and like good wine they are always welcome, and there is a desire for more as they slowly mature. Two such intuitions emerged at this late point in these travels, neither of which disintegrated in a cloud of sparkling raindrops, as happened once earlier. As these later ones matured they left mixed feelings as to their relevance, yet both seemed plausible. It was the conviction that mankind should never ignore opportunities to express concern for its offspring that prompted the recording of these thoughts.

The initial intuition concerned the first recognisable humanoids that were unquestionably warlike. Their lives were short, blighted by infections and injuries. Archaeological data has shown that this direct ancestry extends back little more than 65,000 years, but of particular interest is the fact that the blood group of these nomadic hunters was type O.

Survival for these early forebears depended completely on hunting, and terrible territorial battles repeatedly erupted because of the shortage of vital food. The excessive demands of others were, of course, always considered to be the cause of the warring, which inevitably resulted in many migrations. Eventually domesticated settlements were acknowledged to be a far better alternative than the unsatisfactory feuding and nomadic existence. Time alone was then necessary for substantial communities gradually to develop throughout Africa, Europe, Asia and the Americas.

The blood of these settlers, the descendants of the avid meat-eating hunters of earlier times, remained exclusively type O until around 25,000 BC, and their lives were ideally suited to its characteristics. By the time animal husbandry and agriculture had significantly replaced the nomadic hunting lifestyle, an alternative type A blood group emerged. It could possibly be argued that the evolution of this new blood group was designed to make digestive tracts more tolerant of the grains that were the product of planned agriculture rather than raw or cooked meat. Immune systems able to fight the infections that thrived in densely populated urbanizations were needed. In consequence the mutation from blood type O to type A was very rapid, possibly because a predominance of the type A population survived the many serious exposures to infection.

Type B blood mutated around 15,000 BC and survival rates were complementary to the climate changes within the Eurasian plains. This blood group also tolerated the meat and dairy products that became popular within the more complex

183

communities that developed and in consequence enjoyed robust proliferation.

The only other blood classification of mankind throughout its bumpy ride on planet Earth is type AB. It is a very rare group that emerged less than two thousand years ago. Although well distributed, it is currently found in less than five per cent of Earth's population. However, it is a blood type that has been shown to be special. Its ability to manufacture more specific antibodies to infections has been recognised, as has the fact that the most tolerant immune systems in the modern world have AB blood.

Blood type is determined by two sets of genes at conception. However, as with genes, some blood types dominante others. It is known that type A and type B bloods are dominant over type O. If, at conception, an embryo is given an A gene from the mother and an O gene from the father, the baby will have type A blood, though it will continue to carry the father's O gene in its DNA. When the baby grows up and passes these genes to its offspring, half of the genes will be for type A blood and half for type O blood. Because A and B genes are equally strong, the offspring is type AB if it receives a type A gene from one parent and a type B gene from the other. The type O gene is recessive to all others because it must receive a type O gene from each parent. It is, however, possible for two type A parents to conceive a type O child. This happens when both parents have one type A gene and one type O gene, and each one passes a type O gene on to the child. During conception the relationship is very similar to that of genes for hair and eye colour.

Gene mechanisms are choreographed brilliantly but this discovery trail has taught us to expect such things from the cosmos. Its perfection is unquestionable in every sense of the word. In consequence, its orchestrations should be respected and it is not at all surprising that simple extrapolations show that O type blood is recessive. Quite obviously there is no longer an absolute need for the intensive physical activity that it inspires, or its metabolism, which thrives on the almost exclusively meat diet that was so important for its early nomadic hunting lifestyle. This should in no way be seen as an embarrassment to the massive numbers retaining this blood type. Their preferred protein food is readily available and there are many ways in which those with type O blood can continue to enjoy the vigorous and energetic lifestyle to which they are so well suited. However, it cannot be disputed that planet Earth has become far more environmentally settled than it was twenty to thirty thousand years ago. It was then absolutely essential to kill to eat.

The introduction of type A blood undoubtedly satisfied a need, preparing mankind for the challenges of planned urbanized agriculture. Digestive tracts evolved that suited the agrarian diet, as did personalities that could cope with the changed circumstances and environments. The appearance of B group blood was a similar evolutionary progression to take advantage of the grain and dairy products that were the products of farming. These dairy eaters needed strong immune systems, and the complacent digestive tracts of those with group B blood admirably protected them against the dangers inherent in intensive farm breeding.

Mystifyingly, group AB blood is not at all easy to classify. Undoubtedly this is partly because it is an evolutionary progression of no more than 1,500 years. The fact that it is possessed by only five per cent of the world's population does not help matters. However, there are no uncertainties related to its associated immune system, which is indisputably far-ranging.

How new blood groups are caused to emerge is important and a great deal of data has been published on the subject. Normal evolutionary processes certainly play an important part but it should also be remembered that this journey has taught us that cosmic design is such that evolving universes can never be damaged by the irresponsible behaviour of an organic species. It is therefore not difficult to accept that similar controls exist to protect individual species against adverse natural occurrences.

Blood type is determined by two sets of genes at conception but it is difficult to explain how or why the recent introduction of AB blood could have been genetically driven. Its present distribution suggests that it has arrived in far too many places too quickly for it to have emerged by means of the acknowledged laws of natural evolution.

If we accept that O, A and B blood types evolved in the manner acknowledged, for type AB blood to have arrived in the way it did, cellular genomes would surely have had to be modified at some point in a manner that is inconsistent with natural selection. There had been no noticeable evolutionary cause. It is possible that blood type AB was triggered on planet Earth by the cosmic controls already discussed. An isolated emergency could have been the cause. These were the thoughts that prompted the idea that an extreme situation could have existed. So can any worthwhile conclusions be reached from this reasoning?

It is certainly necessary to acknowledge that the introduction of AB blood at that stage in mankind's evolution is inexplicable in evolutionary terms. There is no obvious environmental cause. An examination of its characteristics and progression, in order to establish exactly why it did emerge, should therefore be encouraged. This may reveal inbuilt safeguards, possibly related to immunity, that could have been the reason for its introduction, though its protection might only have extended to AB individuals. Perhaps scientific analysis of the blood should look for other reasons. It is possible that the protective inbuilt mental safeguards of the cosmos only extend to the behaviour of species. As discussed earlier in this cosmic adventure, having no common language is dauntingly restrictive. So is this a warning for all mankind? Should consideration be given to the possibility that it is being told that immunities within this AB blood could protect everyone against a disastrous pandemic? The recent Asian bird flu scare obviously comes to mind.

Can sufficient understanding of cosmic behaviour be promoted to encourage the complex analytical work required, just in case AB blood carries far ranging immunities that will be desperately needed. It could be of unimaginable importance to the survival of the mass of mankind.

Somewhat apprehensively, an earlier intuition is recalled. At that time it was concluded that to look for keys secreted within mankind's genome structure that unlock total knowingness must stretch mankind's ingenuity as never before. Finding a message expressed in the intuitive language of mental images within the human genome was then the consideration. Compared with that, seeking an explanation for AB blood by researching its make-up should be child's play. But the consequences could be immensely important and unbelievably rewarding.

Intuitions can be looked on as audacious. But it is cosmic behaviour that is being talked about, which surely has to be the ultimate in fearlessness. And it has been said that 'thinking is the valiant effort of the soul'. All that is known for certain that relates to this intuition is:

- **Blood group A rejects blood group B, and B rejects A.**
- **Blood group AB accepts all other blood groups but is rejected by all others.**

- Blood group O can receive only its own blood group. However, it is a universal provider – all other blood groups can receive it.

The second intuition took longer to mature. It was a gradually enveloping feeling that inorganic patterns within maturing planets might be cosmically steered. The objective would be to cause geological shapes to configure in ways that meet the known needs of particular species. The 'will of the cosmos' would, in fact, be favourably adjusting emerging planetary homes to suit the perceived needs of variant species. Such planning would obviously need to take place during the early period of a solar system's evolution.

The introduction of organic life to planet Earth created the need for both plants and animals to receive energy, a problem that was resolved for plant life by photosynthesis, allowing all rooted organisms to use the energy of a sun. This was not a practical proposition for mobile animals, so digestive organs were caused to evolve within their bodies. Eating plant life has been shown to result in the energy collected originally by the plant from the sun being passed on to the animal by the complex activity of eating, digesting and evacuating the plants and their fruit.

One behaviour pattern of this planet is of particular interest because it gives substance to the cosmic persuasion concept under discussion. It began with the introduction of living organisms as distant as three hundred million years ago. It was the period when animals and plants slowly evolved from their simplest forms. At this time in planet Earth's history, organic decay caused vast reserves of coal, oil and gas to be trapped within accessible geological layers of its crust. Thousands of years later, mankind's awareness led to its discovery, as well as to the realisation of its potential. This power source, which owes its origin to the sun, made possible the current industrial age, and the invention of the progressively sophisticated mechanical and related financial tools. These now power the basic activities of planet Earth, where billions of humans live and die.

The energy reserves discussed, though vast, are limited, and in a frighteningly short period of time they will be exhausted. From mankind's point of view it is a very serious problem. Indirectly it could catastrophically affect billions of people. It is indeed true that a great deal of frustration within the many geographical homelands of mankind has already resulted from the half-realised possible consequences. These suggest that fossil fuel power sources will, from all practical points of view, be exhausted within fifty years or so.

Countless technologies are completely dependent on these energy sources, and available alternatives are either hopelessly inadequate or dangerous. A known and tried alternative energy source is nuclear fission. It is the only popularly accepted practical alternative to oil, gas and coal, but regrettably it is extremely dangerous. There are three reasons for this, all of which result from the fact that the energy is released when the structure of the atom is changed during a controlled chain reaction. This process produces energy in the form of heat, but unless handled with great finesse there can be catastrophic releases of uncontrollable radiation. The contamination can cover vast areas of land indefinitely and cause horrendous loss of life. Unsolved problems related to nuclear waste add to these dangers, for it can only be stored and the amounts would be massive should the world be forced to depend completely on atomic fuel for its industrial and pleasure needs. It would then become impossibly dangerous, for no such storage could be safe forever. It is also quite easy for this same fuel to be designed into weapons of mass destruction, which would be at the fingertips of many unstable regimes if nuclear reactors were freely used in the world.

The rights and wrongs of human behaviour are not, however, under consideration. The present objective is to explore the extent to which cosmic intellect might guide planetary designs to the advantage of aware species during their lifecycles. So, can it be argued that cosmic controls could have caused organic fuels to be locked in comparatively easily locatable geological layers of the crust of planet Earth – and those of other solar systems within the cosmos that have similar habitation patterns? Ocean-based aware species such as dolphins would most probably have little or no use for such fuel but their journey towards cosmic adulthood might excite quite different adjustments.

Did cosmic laws contrive planet Earth's reserves of fossil fuel as a transitory driving force for a technologically gifted species on this planet? If so, is there possibly another safe source of power built into cosmic planning for a stable species to use during the remaining five to ten billion years of Earth's useful life? It is extremely difficult to believe that there would not be. If this is the case its early discovery is important, yet the route to this basic need would seem to be beyond mankind's technical ability. Might intuitive interplanetary communion with outer space species be the expected route to achieve this knowingness and could its progression be cosmically guided?

Earlier in this exploration surprise was expressed that mankind never seemed seriously to have envisioned direct communication with the mind of the cosmos. It was considered to be odd that it has not been looked on as an option. That the human genome might have to be researched beyond purely medical needs in order to unlock such a progressive objective, might now be considered imperative. Swiss doctors seem to have unlocked a 'cosmic control' within the brain that restricts organic sight. This may now be of great significance because earlier in this narrative it was suggested that the angular gyrus in the right cortex of the brain possibly includes an organic mechanism that causes out-of-body seeing. Further research into brain function could possibly reveal other trigger points within this vital organ, and could even locate ones that open doors to the cosmic mind itself: its sight, its intellect, its very being.

Could surgeons really use electrodes on selected areas of the brain to open such doors? It may make possible instantaneous intuitions anywhere and everywhere. Under such conditions 'seeing' the full extent of this universe, as well as more distant ones, may be achievable, allowing civilisations advanced beyond mankind's imagination to be studied and viewed. All would then be instantly available for intuitive cognition. Finding an appropriate replacement fuel source would then be a minor problem for suitably adapted species.

Could mankind possibly give serious consideration to the realisation that surgeons and not hazardously propelled space rockets might explore the universe of planet Earth and the 'oneness of mind and purpose' of the cosmic whole?

Eternal concrescence within the omnipresence of the cosmos is a wonderful thought. An internet *par excellence*!

20

Final Thoughts

A NUMBER OF observations gathered throughout this journey within the cosmos remain unstated. Although many have been discarded as unnecessary, others seem to be of importance, though they were not essential to the picture that was being painted. They are now included.

Free Will
It is indisputable that unpredictability is a feature of universes. It is most noticeable in atoms and the signalling protein molecules of living cells. Both have been described as having personality, which ensures variety at inorganic and organic levels. Another way in which it occurs is in the random distribution of genes that take place during the varying reproductive acts of organic life. There is seemingly also the manner in which cosmic enlightenment is 'plucked from the sky' by species journeying towards total conscious awareness. These processes indicate that the universe is not fully predetermined and, therefore, that the concept of free will remains a debatable subject.

Belief and Intuition
There may be those who would argue that journeys such as this cosmic adventure are based purely on personal belief and should

therefore be discouraged. However, the concepts discovered along the way are quite obviously ones that can be personally verified by the reader. The value of their contribution to general knowledge has to depend upon the soundness of the arguments presented. It is regrettable that cosmology cannot show an abundance of absolute proofs – pictures if preferred – in the many ways that other sciences can. It is unfortunate that there is no way experimentally to show that cosmic enlightenment is 'out there' to be absorbed intuitively by consciously aware species. But neither is there a shred of evidence to suggest that brains invent knowledge, whether in the form of inspired mathematical equations or the nuances of any other language.

Proof

Because the cosmos is largely untouchable, and to a great degree unseen, those who grapple with its difficulties in an endeavour to discover its truths have to accept that the burden of proof is delicately balanced. It is this very fact that makes odysseys such as this stimulating. As remarked early on, establishing a balance between mathematical linguistics and intuition is difficult.

Social Justice

It has been shown here that equal opportunity and fairness for all are very desirable social attributes. Democratised societies ostensibly offer such benefits, though wealth continues to be the spur that drives entrepreneurs to success. In consequence massive social differences exist between low-paid workers and successful industrialists. Great actors, highly talented sportsmen and pretty girls can enjoy the same privileges, and one must agree that all contribute to the enjoyment and pleasure of very large numbers of people. However, it is not easy to come to terms with these injustices. Clearly mankind should manage its affairs

193

more sensibly, but knowledge acquired during this journey does not even begin to suggest how. As Isaac Newton ducked the question of the substance of gravity, so this problem of fairness must be left for others to solve. It has been shown that good and evil, right and wrong are meaningless to the cosmos; they are concepts of man's invention. Cosmic visualisations can only be *am, always was* and *always will be.*

Mankind's Gene Count

When these final thoughts were under consideration I was distracted by a Jonathan Mann CNN television programme. He was interviewing Alan Norman of America's well-known *Science* magazine and in the studio three written questions were prominently displayed.

- *What is the universe made of?*
- *What is conscious awareness?*
- *Why have we so few genes?*

I feel confident that readers of these 'Final Thoughts' will have no problem answering the first two questions, and I would like to believe that an inkling of why we have so few genes would also have emerged. In that respect I do, of course, have an advantage. I have given a great deal of thought to this cosmos of ours over the last few years and my recorded adventures have allowed us all to realise that these 'microchips' within our every cell derive from cosmic grain. They are created from this all-knowing grain of thought and substance.

As species advance to improved levels of awareness as a result of direct cosmic intervention, considerable numbers of the genes in their cells become redundant. This is simply because their purpose can be achieved more efficiently by the direct action of cosmic grain rather than its fabricated genetic tools of the trade.

There is no way in which I can suggest what the final requirement of genes will be within species which have reached the maximum possible level of conscious enlightenment. I feel confident that the requirement of genetic knowingness within the cells of organic life will always be substantial, though it will be considerably less than at present.

Lynne McTaggart's Book *The Field*

It was received with a note – 'Ken they're getting close' hastily scrawled on the cover. It was a most pleasing and much enjoyed gift from the golfing friend who never received the promised letter that eventually became this book.

Reading *The Field* excited intuitions related to this cosmic journey, the finished manuscript of which was already in the hands of its publishers. Memories emerged of an inherent cosmic intellect that acknowledged the operation of chance. I recalled the conclusion that the life of our planet is not sufficiently extended for such patterns to have significance, and also the later insight that the eternal, ubiquitous cosmos exhibits eternal behaviour patterns so that every possible activity will be repeated endlessly. This has to be so. Mathematicians universally accept that tapping away randomly on a typewriter must eventually reproduce all Shakespeare's plays word perfectly if time is not of the essence.

The cosmos stands the test of time. Its universes, planets and organic life are infinitely diverse; because there is endless time their forms and characteristics must be determined by the law of random tapping on a typewriter. Every possible combination will be repeated continually down to the minutest detail. There is no possible combination of events or circumstance that has not already been and will not continue to be enacted repeatedly within the all-pervading timelessness of cosmic reality.

Not only is cosmic grain ubiquitous, but every creation and every activity is repeated ceaselessly and unendingly – and that includes *us*. Consciously aware individuals must decide for themselves what satisfaction this awesome fact holds for them.

A FINAL WORD

It is said that every picture paints a thousand words.

These intuitive 'pictures within' are the language of the undetectable grain of cosmic reality, which is intellectually complete. It is an indestructible and eternal omnipresence.

Mankind is part of this indestructible intellect, the grain of which is eternally the essence of everything.

Lord Byron wrote:

> *But words are things, and a small drop of ink*
> *Falling like dew upon a thought, produces*
> *That which makes thousands, perhaps millions, think.*

Index